COLUMBIA COLLEGE CHICAGO

3 2711 00087 6577

D1786758

DATE DUE

MAY 0 8 2013			
			OCT 1 9 2005

DISCARD

GAYLORD — PRINTED IN U.S.A.

Changing Concepts of Time

CRITICAL MEDIA STUDIES
INSTITUTIONS, POLITICS, AND CULTURE

Series Editor
Andrew Calabrese, University of Colorado

Advisory Board

Patricia Aufderheide, American University
Jean-Claude Burgelman, Free University of Brussels
Simone Chambers, University of Toronto
Nicholas Garnham, University of Westminster
Hanno Hardt, University of Iowa
Gay Hawkins, The University of New South Wales
Maria Heller, Eötvös Loránd University
Robert Horwitz, University of California at San Diego
Douglas Kellner, University of California at Los Angeles
Gary Marx, Massachusetts Institute of Technology
Toby Miller, New York University
Vincent Mosco, Queen's University
Janice Peck, University of Colorado
Manjunath Pendakur, Southern Illinois University
Arvind Rajagopal, New York University
Giuseppe Richeri, Università Svizzera Italiana
Kevin Robins, Goldsmiths College
Saskia Sassen, University of Chicago
Colin Sparks, University of Westminster
Slavko Splichal, University of Ljubljana
Thomas Streeter, University of Vermont
Liesbet van Zoonen, University of Amsterdam
Janet Wasko, University of Oregon

Recent Titles in the Series
Digital Disability: The Social Construction of Disability in New Media,
    Gerard Goggin and Christopher Newell
Principles of Publicity and Press Freedom,
    Slavko Splichal
Internet Governance in Transition: Who Is the Master of This Domain?
    Daniel J. Paré
Recovering a Public Vision for Public Television,
    Glenda R. Balas
Reality TV: The Work of Being Watched,
    Mark Andrejevic
Contesting Media Power: Alternative Media in a Networked World,
    edited by Nick Couldry and James Curran
Herbert Schiller,
    Richard Maxwell
Harold Innis,
    Paul Heyer
Toward a Political Economy of Culture: Capitalism and Communication in the Twenty-First Century,
    edited by Andrew Calabrese and Colin Sparks
Many Voices, One World,
    The MacBride Commission
Changing Concepts of Time,
    Harold A. Innis

Forthcoming in the Series
Public Service Broadcasting in Italy,
    Cinzia Padovani
Film Industries and Cultures in Transition,
    Dina Iordanova
Globalizing Political Communication,
    Gerald Sussman
The Blame Game: Why Television Is Not Our Fault,
    Eileen R. Meehan
Mass Communication and Social Thought,
    edited by John Durham Peters and Peter Simonson
Entertaining the Citizen: When Politics and Popular Culture Converge,
    Liesbet van Zoonen
Elusive Autonomy: Brazilian Communications Policy,
    Sergio Euclides de Souza
Raymond Williams
    Alan O'Connor

Changing Concepts of Time

Harold A. Innis

Introduction by James W. Carey

ROWMAN & LITTLEFIELD PUBLISHERS, INC.
Lanham • Boulder • New York • Toronto • Oxford

ROWMAN & LITTLEFIELD PUBLISHERS, INC.

Published in the United States of America
by Rowman & Littlefield Publishers, Inc.
A wholly owned subsidiary of The Rowman & Littlefield Publishing Group, Inc.
4501 Forbes Boulevard, Suite 200, Lanham, Maryland 20706
www.rowmanlittlefield.com

P.O. Box 317, Oxford, OX2 9RU, United Kingdom

Copyright © 2004 by Rowman & Littlefield Publishers, Inc.

All rights reserved. No part of this publication may be reproduced, stored in a retrieval system, or transmitted in any form or by any means, electronic, mechanical, photocopying, recording, or otherwise, without the prior permission of the publisher.

British Library Cataloguing in Publication Information Available

Library of Congress Cataloging-in-Publication Data
Innis, Harold Adams, 1894–1952.
  Changing concepts of time / Harold A. Innis.
    p. cm. — (Critical media studies)
Includes bibliographical references and index.
  ISBN 0-7425-2817-0 (cloth : alk. paper) — ISBN 0-7425-2818-9 (pbk. : alk. paper)
 1. Mass media—Political aspects—United States. 2. Mass media and culture—United States. 3. Canada—Civilization—American influences. 4. Communication. 5. Space and time. 6. Economics. I. Title. II. Series.
 P95.82 .U6I54 2004
 302.23'0973—dc22
                                                                    2003021044

Printed in the United States of America

∞^{TM} The paper used in this publication meets the minimum requirements of American National Standard for Information Sciences—Permanence of Paper for Printed Library Materials, ANSI/NISO Z39.48-1992.

Contents

Introduction to the Rowman & Littlefield Edition *James W. Carey*	vii
Editor's Note	xxi
Changing Concepts of Time	
Preface	xxv
1 The Strategy of Culture	1
2 The Military Implications of the American Constitution	21
3 Roman Law and the British Empire	45
4 The Press, a Neglected Factor in the Economic History of the Twentieth Century	73
5 Great Britain, the United States, and Canada	105
Index	129
About the Author	133

Introduction to the Rowman & Littlefield Edition

James W. Carey

> The "winner" of the Cold War will inevitably face the *imperial* problem of using power in global terms but from one particular context of authority, so preponderant and established and unchallenged that its world rule would almost certainly violate basic standards of justice.
>
> —Reinhold Niebuhr, *The Irony of American History* (1952)

In the early summer of 1952 Harold Innis left the hospital where he was being treated for terminal cancer. In the last months of his life, while at home, he edited the page proofs for *Changing Concepts of Time*, the final manuscript he would deliver to the printer. In early fall, the disease broke through his remaining defenses and he died on November 8, 1952.

His illness and death ended four years of intense and productive scholarship undertaken in extreme circumstances. Between 1948 and 1952, in an astonishing burst of creative energy, he wrote virtually all the work for which we today celebrate him as the major theorist and historian of communications in North America. The books that had brought him fame as an economic historian, geographer, and theorist—*The Fur Trade in Canada, The History of the Canadian Pacific Railroad, The Cod Fisheries*—are little read today except by biographers and economic specialists and are largely out of print. However, his works on communications—*Empire and Communications, The Bias of Communications, The Press: A Neglected Factor in the Economic History of the Twentieth Century*, and now *Changing Concepts*—are still available and widely influential today.

Unlike his earlier books on economics, the volumes on communications were exclusively comprised of essays, loosely mortised together, that had originally been delivered as lectures. In the spring of 1948 alone he delivered the Beit Lectures at All Souls College, Oxford, which became *Empire and Communications*, and, on the same trip to Great Britain, the final two essays of this book were delivered as the Stamp Lecture at the University of London and the Cust Lecture at the University of Nottingham.

The essay form reflected not only the tentativeness of his thinking but also the urgency of his task to, as an even more famous political economist put it, not only analyze the world but change it. He wryly commented that "I once had to choose between going into university work or politics and I decided to go into politics" (Watson, 1977, p. 45). The cryptic nature, intensity, and energy of his communications essays reflect not only his struggle against physical decline but also the extraordinary burdens he had assumed. While head of the Department of Political Economy at the University of Toronto (a post he had held since 1937), he was also concurrently dean of the Graduate School. While serving a term as president of the Royal Society of Canada, he simultaneously traveled the country and conducted hearings for the Royal Commission on Transportation, of which he was a member.

Innis was one of those thinkers (Marx was another) whose work is regularly divided into early and late phases, separated by a "gap" marking an abrupt change of subject matter and philosophical outlook. The suddenness with which the work on communications appeared, and the contrast between the bold theorizing of the later work and the precise if sometimes numbing detail of his studies in economics, reinforces the belief in a breach between the "young" Innis and the mature scholar. I do not believe such a separation exists; there is no radical disjunction between the early and the late Innis. His work on communications naturally grows out of and elaborates upon his early studies of the economic and political formation of Canada, as the essays in this volume attest. His subject was always empire, globalization, international trade, "the *longue durée* of events and epoch-making forces that transformed economies, states and civilizations" (Drache, 1995, p. xix). Innis was an economist of trade rather than production, of routes, movements, flows, and circulations rather than of factories and modes of production. The international economy of trade was powered, in his view, by progressively improving means of transportation and communications, parallel interacting systems of economic and social expansion and consolidation. This was as true of the expansion of the European trading system into North America in the seventeenth century, and the growth of imperialism in Africa and Asia in the nineteenth, as it is today in the age of jet aircraft and the Internet. The contradictory and unintended consequences of technology, when linked to the equally contradictory but thoroughly human desires for economic, political, and cultural domination, provided him with a lifelong subject matter.

In the years after World War II, the international system—which had lain dormant in the carnage of two world wars—reawakened in a radically altered technological and political setting. The year bridging 1948 and 1949 was pivotal both in politics and in Innis's studies, and that is where this particular story begins.

On February 8, 1949, an innocently named "Values Discussion Group" was convened at the University of Toronto. It was chaired by economist Thomas Easterbrook and included as members not only Harold Innis but also an instructor of literature from St. Michael's College, Marshall McLuhan, who had joined the faculty three years earlier. They met weekly during the semester with each member of the group in turn presenting a paper and leading the discussion.

Such discussion groups, a hardy perennial of the academic garden, flowered across the continent at this moment. However named, they shared a common purpose: to deal with the widely perceived "crisis of civilization" of the postwar period. "Crisis" is one of the most widely and idly used terms of the civic vocabulary, and Innis warned against it: "It will not do to join the great chorus of those who create a crisis by saying there is a crisis." But the barbarism of the twentieth century, which was fully revealed only when the war ended, and the drift from alliance to cold war that threatened the peace seemed to warrant the word crisis.

All that remains in the archives of the University of Toronto are the minutes of the meetings of the "Values Discussion Group." Judging from them, most of the gatherings were pretty dreary, focusing on the role of values in scientific research: the so-called fact-value dichotomy.

Marshall McLuhan made the fifth presentation to the group, and that too was pretty tame—a plea for art as against science—particularly considering the intellectual pyrotechnics he set off two decades later. McLuhan was still in his "Mechanical Bride" phase lamenting the decline of literature in an age of mass culture. He was then in the grip of the American cohort of "New Critics," particularly those like Allan Tate and J. C. Ramson, who recruited him into their rearguard campaign against, in Tate's phrase, the "all destroying abstraction of [industrial] America." McLuhan looked to the American South as an outlet for his preindustrial yearnings. The South stood as a living, thriving monument of the Pastoral Ideal—if not earthen cottages, clotted cream, and the God-fearing peasant of English romantic poetry—then at least a brackish sanctuary from capitalism and its individualism. The region nurtured a distinct cluster of values that McLuhan, like others, called the "Southern Quality": an aristocratic humanism, an agrarian economy, and worship of the cyclical rhythms of the land. The South was heir to an alternative cultural tradition that "took its stand" against the spiritless rationalism of the North. The arts—those McLuhan admired, at least—were a storehouse of values. He lionized a period when the arts were not separated from life as they had become in the modern world. He elevated Art and the Artist, now capitalized into cultural exemplars, to the role of explorer and innovator, restless seekers of new continents of meaning who could not tolerate the banal, ordinary, standardized, repetitive, and routine—the archetypal features of industrial civilization.

When Harold Innis rose to speak at the eighth and last meeting of the group, the trivial and romantic aura that had dominated earlier sessions evaporated. One cannot help but infer from the notes that he believed a genuine crisis was in tow, and its nature was not to be found in technical philosophy or romantic poetry. The crisis was one of politics, economics, and communications.

As World War II came to an end, two very large problems loomed on the horizon. The first was the fear that both national and international economies would slip back into the economic nightmare of the 1930s and replay the hostilities that had recently ended. Millions of veterans in all the warring powers had to be absorbed into the civilian economy and given productive work. Unemployed veterans wandering the streets were a recipe for civic disaster. What would happen as defense spending slowed and factory jobs disappeared? Would consumer demand be strong enough to offset the drop in military production? Would it be possible to govern such societies? And if the route of creating a consumer society, the one eventually followed, were taken, what would happen to the political culture of the 1930s in which citizens organized to protect both the individual and social interest through boycotts, publicity campaigns to oppose child labor, demands for pure food and drug laws, and support for the rights of workers to join unions? Would passive consumers replace active citizens as the necessary price of economic recovery?

The second problem was the outbreak of the Cold War and the nuclear arms race. Innis had traveled through the Soviet Union at the end of the war and had a premonition of the conflict that was to follow as two great empires—Soviet and American—emerged to organize the political world. The phrase "Cold War" was coined in 1948 by Bernard Baruch and in July, as Innis edited *Changing Concepts*, the Berlin airlift, with all its associated dangers, began. Canada in particular was in a difficult situation. It had long been a colonial outpost dominated by Great Britain from which it had inherited its basic institutions and culture. Now Canada was trapped between two modern empires, one of which was on its doorstep. How was Canada to preserve its political and cultural independence and remain true to its British origins? How could it remain politically erect, part of a third bloc, given the pressure, influence, and proximity of the United States? What constructive role could Canada play in the violent and unstable world that loomed ahead?

Moreover, technological innovation, held back and redirected to the war effort for most of the decade, was reemerging as the engine of economic recovery and military competition. By 1948, television had started its relentless colonization of politics and culture as it spread from domestic capitals in both the United States and Canada into the hinterland. In the same year, Norbert Weiner's (1948) landmark book *Cybernetics: Or, Control and Communication in Animals and Machines*, which summarized decades of research on self-organizing systems, argued that electronic servomechanisms were the technological twins of television and the instrument for the automation of work and knowledge. Weiner suggested that the purpose of communication was to control the environment, but in order to communicate effectively it was essential to consider feedback as the mechanism governing the sending of messages. To govern, whatever the object—animal, human, or natural processes—requires one to consider the audience (or the receiver or destination in cybernetic terms) in order to alter the message relative to the feedback. But was cybernetic governance, understood as a form of control rather than a mode of participation, in opposition to democratic politics? What were the social consequences of conceiving communication as a control mechanism within a feedback loop? Weiner's question, posed later, was one Innis posed in a different vocabulary: What, in the age of communication and control, are the "human uses of human beings"?

Innis began by arguing that increased savagery followed developments in communications and transportation. New techniques upset old values before standards could be developed to control the technology. The development of printing, with its emphasis on nationalism and the vernacular, had set loose wars of national liberation that overthrew the Holy Roman Empire and gave rise to the modern international system. Similarly, the emergence of the telegraph and telephone, wedded to high-speed oceangoing navigation, initiated the imperial competition among the European states, climaxing in World War I. Modern warfare illustrated that the truism held for electronics, radio, and aviation as well.

Innis is among the earliest and most trenchant theorists of globalization in both its economic and communications dimensions. His globe was a more limited one than ours, confined in his major work to the European Atlantic corridors to the New World, but it was the global system of communication and culture, always in relation to Canada, that was his central concern. His early work in economics concerned the creation of North America—the actual shaping of the land and the plantation of foreign cultures—as an outpost of European empires: New France, New England, New Amsterdam, and Nuevo Espanola. While Europeans came to North America for many reasons—in search of religious freedom, to found new communities—the overwhelming impulse was the exploitation of the commodities of the region: cotton, tobacco, cod, fur, and gold. To affect this exploitation Europe needed both a reliable commercial communications system and an actual cultural plantation. Both were made possible by the variable capacities of oceangoing navigation and literacy: the ability to move through the "cultureless" void of the ocean without the contaminating effects of human contact and to connect and coordinate imperial outposts via news, newsletters, and the printing press. For the first time in history, the Atlantic shipping lanes carried the furniture of entire cultures in one direction, and, in the other, the natural products of North America transformed into commodities by the demand of nascent capitalist markets.

This first phase of globalization ended in the late eighteenth and early nineteenth centuries when Europe outran its lines of communications and was unable to maintain, in his phrase, a monopoly of knowledge and force. Europe's capacity to dominate space through communication, markets, law, and force was eroded at its most distant margins where alternative cultures, cultures of Creole nationalism, grew up. In the Americas, Europeans were transformed; they found a new identity and, in the practical tasks of adapting to a new environment, created forms of knowledge and self-understanding—a new culture—radically different from the one carried on the voyages of exploration and settlement. Creole nationalism grew in North America along the geographic fault lines implanted by the imperial powers: at the margins of such powers (along the St. Lawrence River, for example) or along lines carved for the administrative convenience of Europe such as in Spanish America.

Innis did not have the contempt for empire typical of today, when even imperial peoples protest their innocence. Empires could be good or bad, republican or authoritarian, benign or destructive, progressive or reactionary. To inveigh against empire was to tilt against a windmill, for empire is a persistent form of social organization, one practically as old as our knowledge of human history. The intellectual problem was one of understanding the conditions under which empires were created and dissolved and the standards by which to judge their effectiveness and civilizing potential.

After the first phase of globalization, European empires were redirected toward other continents via newer and more rapid forms of travel and communication—steam and electricity. At the same time, new empires grew in the Americas as nations pursued their own manifest destiny, seeking to expand over neighboring landmasses, in some cases stretching from the Atlantic to the Pacific. Prior to World War I there were two kinds of empires: landed empires, products of centuries long expansion over contiguous territories that the United States and Canada imitated in quite different ways; and overseas colonial realms. "Among the first group—Russia, Austria-Hungary, the Ottoman domain, China—the states *were* empires and were vulnerable to new forces of national self-determination. Members of the second group—the British, French, Dutch, Spanish and Portuguese . . . —*had* empires. When the internal crises of the first group combined with the interlocking rivalries of the second, the result was the First World War" (Maier, 2002, p. 29). The history of the twentieth century can be viewed as one interlinked history of imperialism—"from the domination and then the destructive rivalries of the Europeans, to the Soviet and American spheres of influence that emerged from the Second World War and finally to the ascendancy of the United States as the only remaining superpower" (Maier, 2002, p. 29). Empire building, whether by landed expansion or in overseas colonial realms, was the dubious achievement of the nineteenth century. But that phase of globalization ended when the guns of August sounded in 1914. The years from 1914 to 1948 marked an interregnum in the international system—marked by a severe and nearly universal economic depression and two great wars that ended in a Cold War and nuclear standoff. States were absorbed in warfare at home, and sometimes in the colonies (the first shot of World War I actually came off the coast of Australia), and with it all the other hallmarks of international movement—immigration, capital flows, and trade—declined. International trade and capital flows would remain below 1913 levels until the mid-1970s. Immigration measured as a proportion of world population has never fully recovered.

Innis's complex histories of trade, commodities, technology, and communications largely examined the first two phases of globalization: (1) the colonial settlement and expansion of North America and (2) the nineteenth-century imperial competition to control distant territories. In 1948, the long parenthesis that had enclosed the period from the opening of World War I to the closing of World War II was about to be breached. How was it to be breached? Two possibilities existed: a global conflict and struggle for power between the East and West, the Soviet Union and the United States, or the replacement of rivalries, old and new, by institutions of collective security and cooperation. To appraise these questions he needed, in addition to the bag of concepts acquired in economics—concepts such as monopoly, equilibrium, unused capacity, liquidity preference, and market structure—two things: a way to apply these concepts more systematically to the phenomena of communications and, even more desperately, a moral and ethical counterbalance to the bias of modern civilization.

It is an autism of Western scholars, as natural as a plant turning toward the light, that in moments of crisis one turns to the foundation of Western civilization, to Greek mythology and philosophy, as a source of renewal. So, Innis turned to the field of classical studies in "which he mispronounced the names of even the most common authorities" (Watson, 1977, p. 45). He was aided by the fact that the University of Toronto had a splendid Department of Classics and within the department a great student of Greek thought, Eric Havelock. Havelock and Innis worked independently and only discovered one another four years after Havelock left Toronto for Harvard. Havelock's book, *Prometheus Bound: The Crucifixion of Intellectual Man* (1950) is, as Innis acknowledges in his preface, the persistent background that controls *Changing Concepts of Time*. Both Innis and Havelock were puzzling through the relation between intelligence and power. Havelock interpreted the Prometheus myth as a drama symbolizing the conflict between the short-range intelligence of the political class in pursuit of power, as represented by Zeus, and the long-range intelligence of the intellectual class in pursuit of understanding as represented by Prometheus the Forethinker. Havelock applied the myth to the events of World War I and its aftermath. Both Innis and Havelock believed that both forms of intelligence were necessary in human affairs, for the short-run intelligence of power represents the problems of space and control while the long-range intelligence of science represents time and the spirit of conversation, dialectic, and compromise. During World Wars I and II, intelligence deserted its proper role and embraced and served power in what Julien Benda had earlier called "The Great Betrayal." As Havelock comments:

> The first struggle of the twentieth century, between 1914 and 1918 . . . became a war of factories, cities and total manpower. It was fought with an abandonment of political compromise, by the elimination of all avenues of agreement, by the refusal to consider truce or armistice. The objectives became unlimited, the peace . . . was framed to express a complex of hates and fears and revenges, in which scientific calculation . . . had no part [and] the agreements were little founded in history, in sociology, in economics, or even in physics and chemistry. (Havelock, 1950, p. 26)

In other words, those in political power exercised a monopoly of knowledge over the public domain. They were exclusively present-minded, seeking the satisfaction of their own interests, driven by shortsighted hatred and desires for revenge that they systematically implanted and exploited in public discourse. Power was indifferent to the long run and the larger interests of humankind. The voice of the scholar was silenced or, even worse, co-opted by power into a tool of the state. This monopoly of knowledge was founded on the media of print and broadcast, which reinforced the tendency to live exclusively in the present, in a world defined by the news cycle: the day or increasingly the hour or quarter-hour. We were kept waiting for the news as a substitute for participation in politics. The temporal horizon collapsed into the present, and forethought, planning for the future, thinking in terms of posterity, became obsolete.

In *Changing Concepts of Time*, Innis is no longer attempting to elaborate a theory of time, space, and media. The reader will search in vain for an answer to the question posed in the title: What are the changing concepts of time? Instead, he says, the essays represent an attempt to apply "the thesis developed in *The Bias of Communication* and *Empire and Communication* to immediate problems." What is that thesis? The spatial bias of modern media, the attempt to extend lines of communication further and further, from center to margin, from the capital to the hinterland, in order to exercise definitive control over the environment, including the humans that inhabit that environment, inevitably shrinks time down to the present, to a one-day world of the immediate and the transitory. The future disappears into the present; everything changes at a blinding speed, making it difficult to maintain continuity in time and culture. What is the overriding problem? The current crisis, he says, thinking of the Cold War, is "the intellectual organization of political hatreds." He attempts to illustrate throughout these pieces one of his favorite maxims: the more the technology of communication improves, the more difficult human communication becomes. "The problems of understanding others have become exceedingly complex partly as a result of improved communication," for the problem of understanding recedes in the face of the insistent emphasis on the present and the exercise of domination. He concludes:

> The present—real, insistent, complex and treated as an independent system—has penetrated the most vulnerable areas of public policy. . . . War has become the result and a cause of the limitations placed on the forethinker. Power, and its assistant force, the natural enemies of intelligence, have become more serious as "the mental processes activated in the pursuit and consolidating of power are essentially short range." (Innis, 1952, pp. v–vi)

Changing Concepts is essentially an extended essay on a variety of problems that revolve around the emergence of United States as the successor to Great Britain, as the dominant modern empire. "American imperialism has replaced and exploited British imperialism," and Canada has moved "from colony to nation to colony." Empires rule not only by force and power but also, and perhaps more importantly, by exercising monopolies of knowledge, controlling not only routes of trade but routes of culture: artistic styles, language, consumer preferences, and intellectual ideas.

But American imperialism, as opposed to British, presents certain unique problems. For one thing, American culture is relentlessly opposed to tradition because the central elements in it are the product of and supported by the system of mechanized communication devoted to a systematic, ruthless destruction of the elements of permanence essential to cultural activity. "The jackals of communication systems are constantly on the alert to destroy every vestige of sentiment toward Great Britain holding it of no advantage if it threatens the omnipotence of American commercialism." Canada faced a postwar onslaught of a resurgent American cultural industry led by film and broadcast but supplemented by renewed efforts of American publishers to dominate the "Canadian market." The emphasis on change is the only permanent characteristic of American commercial culture. Every movie, every broadcast program, or issue of a newspaper or magazine must be quickly forgotten, rendered obsolete, in order to clear the way for the next film or program or publication, each of which is unique, unprecedented, unparalleled, extraordinary, exceptional—even though indistinguishable—from what has come before.

American culture is not only without permanence, caught in a process of ceaseless change, but, in a vast country, it is dominated by one location—New York, the home of the communications industries. As a result, "Canadian writers must adapt to American standards. Our poets and painters are reduced to the status of sandwich men."

The power of American culture is reinforced by three underlying tendencies. First, because British books were not afforded copyright protection unless first published in the United States, the number of British books outpaced American books until late in the nineteenth century. As a result, American writers turned to journalism and became expert on that more fragile, less permanent form of publication. Second, the First Amendment was taken to be not only a part of the American Constitution but a universal right to be imposed on others according to American meanings. Innis interpreted the meaning of freedom in the First Amendment to be decidedly parochial, yet it was used as a weapon to grant and guarantee a monopoly to the American periodical press, enforceable without regard to national boundaries and traditions. Many American journalists, for example, have simply ignored Canadian press law when reporting from Canada because they consider such law an offense to the First Amendment. The First Amendment was also used to extirpate freedom of speech as it was aimed at suppressing the oral tradition, turning readers and viewers into passive spectators who did not speak to one another. And third, American economic and cultural policy intensified divisions within Canada, driving a wedge between those provinces dependent on the American market and those dependent on Europe and between the English-speaking provinces and Quebec.

Innis describes the first chapter in this book, "The Strategy of Culture," as a footnote to the Massey Commission. That commission, formally known as the Royal Commission on the National Development in Arts, Sciences, and Letters, argued for the protection and development of an indigenous Canadian culture against the commercialism of the United States. The view of culture espoused by the commission, descended from Matthew Arnold's *Culture and Anarchy*, is essentially conservative and nationalist, while antipopulist in its support of a largely British tradition of high culture. Innis's support of tendencies in the report is something of a surprise, for he was generally "opposed to nationalism as a program or an ideology and even strongly opposed to the exclusivist and intolerant spirit which that doctrine usually incorporated."

He was led to this extreme position by the belief that in Canada "we are fighting for our lives" and "the pernicious influence of American advertising reflected especially in the periodical press and the powerful persistent impact of commercialism have been evident in all the ramifications of Canadian life. . . . The effects of these developments on Canadian culture have been disastrous. Indeed they threaten Canadian national life."

While Innis affirmed Canada's peripheral location relative to the United States and Europe, he insisted that to be a Canadian is not parochial. This was not a narrow, inward-turning nationalism. Instead, he was asserting that Canadians were in possession of a valuable body of experience—a genuine culture of knowledge and understanding—developed in the complex process of adapting to a particular geography and history. In his opposition to monopolies of knowledge and culture, he affirmed a central principle of John Dewey: left to their own devices, humans multiply cultures like the unforced flowers of spring. The value of these cultures resides in their sheer variety—the alternative forms of adaptation, knowing, and understanding they contain. But to sustain this variety against the "conservative power of monopolies" compels the development of "technological revolutions in the media of communication in marginal areas." For Canada to resist requires government involvement in the development of forms of communications resistant to the lures of American commercialism. Only the state can create conditions where cultural production can flourish. The conservative side of Innis is demonstrated by his conviction that a cultural heritage is a more enduring foundation for national prestige than political power or commercial gain. This same conclusion was reached at about the same time by, surprisingly, the Hutchins Commission on Freedom of the Press in the United States.

The postwar period created not only problems surrounding the transnational spread of commercial culture, but, more seriously, problems of military power as well. Throughout these essays Innis reflects on the military impulses within American society and their implications for Canada and the world. Canada has always been faced with the problem of absorption as the fifty-first American state. In the late 1940s, Canada was faced with prospects of being drawn into the Cold War or, worse, into a variety of military adventures.

In chapter 2, Innis argues that certain basic weaknesses of written constitutions in general, the American one in particular, have been exacerbated by "improvements in communication." Because America was founded in violent revolution, it is prone, he believes, to excessive nationalism and patriotism. Founded as a mass democracy, at least by the standards of the eighteenth century, the United States had built into the Constitution protections for minorities against the enthusiasms of the crowd, but these had been considerably weakened by improvements in the technology of communications, which favored direct demagogic appeals to national sentiments. These same improvements favored the presidency over the legislature, thus upsetting the balance of powers among the branches of government and rendering the executive an imperial office.

The two fundamental weaknesses in the Constitution in his view resided, first, in the clauses that make the president the commander in chief of the armed forces with almost unchecked control over foreign policy. While the president must secure consent from Congress for military adventures, once he has achieved it—on however flimsy a basis—he can virtually act as a monarch. Second, by fixing the dates of elections—their occurrence is predictable, unlike those in a parliamentary system—foreign policy is sacrificed to political campaigning. Foreign policy and foreign adventure are calculated in terms of their effect on the next election. This intensifies the obsession with the present and reinforces the willingness to sacrifice long-run political stability to short-run electoral gains. This results in vacillating, inconsistent policy at the mercy of strategy for winning elections rather than the national interest. He comments: "An attempt under the second Roosevelt to establish a bi-partisan foreign policy has given greater stability but foreign issues are all too apt to be dominated by the immediate exigencies of party politics."

Under these circumstances, military domination of foreign policy is inevitable. "The limitations of American foreign policy are largely a result of its lack of tradition and continuity and its consequent emphasis on displays of military strength." The propensity of American politics to elevate military generals to positions of civilian leadership contributes to the power of a single person, backed by constitutional authority, to intervene in war despite the will of Congress. What he observed in the opening of the Cold War—think of Senator Joseph McCarthy—were attempts to systematically arouse public opinion, to keep it constantly agitated concerning the need for war even against the most modest of enemies. He compares this to the fanatic fear of mice shown by elephants.

Innis pinned his hopes for Canadian autonomy during the Cold War to the development of a third bloc independent of the two great rival powers, surely a tough thing to pull off in North America. He doubted that the United States could work though consensual principles of politics shared with its allies, for like all empires, the first goal of the American one had to be establishing and stabilizing a periphery through military effort. Empire builders yearn for stability, but what imperial systems find hard to stabilize is precisely their frontiers. Empires are constantly drawn to expansion by the disorder seething just outside the last domain they have stabilized, but each stabilized zone generates a further zone of chaos that requires imperial policymakers to intervene anew. The use of force that stabilizes conditions within any given boundary often upsets a precarious peace among the tribes or weakened states that abut the frontier. Thus, empires must inevitably generate a resistance that rulers will perceive as shortsighted, bloody minded, and fanatic.

I have deliberately cast this interpretation of Innis in the present tense in order to suggest, however indirectly, that he continues to speak to issues of politics, communications, and empire as this book goes to press and we contemplate the meaning of the first military skirmishes following the settlement of the Cold War. Instructive also in this regard, as a new generation of the "best and brightest" seizes the reins of politics, are words he was fond of quoting:

> I think that the greatest hindrance to constructive political action in the last thirty years has been the influence on final decisions of experts, especially of experts obsessed with the belief that their own generation has gained a vantage

point unprecedented in history. No quality is more important in a political leader than awareness of the accumulated wisdom and experience handed down not only in written documents but also by word of mouth from generation to generation in practical diplomatic, administrative and legislative work. The more we work with mass statistics and large schemes the more we are in danger of neglecting the dignity and value of the human individual and losing sight of life as a whole. (University of Chicago, Committee on Social Thought, *Works of the Mind*, pp. 116–17)

It is impossible to agree with all the arguments concerning politics, culture, and communications that Harold Innis put forth, and, it must be admitted, his work, in the small window of life he had left to execute it, sometimes took on an exaggerated, even hysterical tone. Nonetheless, this is serious, even exemplary, scholarship in its determined attempt to work on a broad canvas and to integrate history and theory with practical understanding. He understood the study of media, as we now call them, not as some walled-off department of thought, but as a pathway to plunge one into the deepest, most intractable problems of contemporary life.

References

Drache, Daniel. 1995. "Introduction" in *Staples, Markets and Cultural Change, Selected Essays [of] Harold Innis*, pp. xiii–lix. Montreal and Kingston: McGill and Queens University Press.

Havelock, E. A. 1950. *The Crucifixion of Intellectual Man*. Boston: Beacon Press.

Innis, Harold A. 1952. *Changing Concepts of Time*. Toronto: University of Toronto Press.

Maier, Charles A. 2002. "An American Empire?" *Harvard Magazine* (December): 26–31.

Niebuhr, Reinhold. 1952. *The Irony of American History*. New York: Scribners.

University of Chicago, Committee on Social Thought. 1947. *The Works of the Mind*, by Mortimer J. Adler et al. Edited by Robert B. Heywood. Chicago: University of Chicago Press.

Watson, A. John. 1977. "Harold Innis and Classical Scholarship." *Journal of Canadian Studies* 12, no. 5: 45–61.

Weiner, Norbert. 1948. *Cybernetics: Or, Control and Communication in Animals and Machines*. Cambridge, Mass.: MIT Press.

Editor's Note

To preserve the original text, we have made few alterations to *Changing Concepts of Time*. As a result, reference information that was not included in the original book has not been added to this edition. For example, publisher names are missing throughout most of the notes in the book. Although this information is pertinent, readers wishing to find the books mentioned by Innis should still be able to do so. Also, sources for quotations are often unattributed and we have chosen to leave them as they are; it would be virtually impossible to trace Innis's sources for these statements. Our goal is to preserve, as best we can, this classic work—the last publication by Harold Innis.

Changing Concepts of Time

To

D. G. C.

L. S. B. C.

Preface

An attempt is made in this volume to elaborate the thesis developed in *The Bias of Communication* (Toronto, 1951) and *Empire and Communications* (Oxford, 1950) in relation to immediate problems. For that reason, unfortunately, it reflects more sharply the temper of the period. The first two essays were published in pamphlet form earlier in 1952 under the title, *The Strategy of Culture*. The other essays wer printed as lectures and have been revised. I am grateful to the sponsors, to whom specific reference is made in each essay, for permission to reprint them.

It has been assumed that different civilizations regard the concepts of space and time in different ways and that even the same civilization, for example that of the West since the invention of printing, differs widely in attitude at different periods and in different areas. Even within a given political region such as the United States the attitude toward time and space will vary in different areas—notably the east and the west. Indeed the political boundaries and the character of political institutions will reflect the variations in themselves. In an attempt to explain these differences emphasis has been given to technological changes in communication. The problems of understanding others have become exceedingly complex partly as a result of improved communications.

The general argument has been powerfully developed in the *Prometheus Bound* of Aeschylus as outlined by E. A. Havelock in *The Crucifixion of Intellectual Man* (Boston, 1951). Intellectual man of the nineteenth century was the first to estimate absolute nullity in time. The present—real, insistent, complex, and treated as an independent system, the foreshortening of practical prevision in the field of human action, has penetrated the most vulnerable areas of public policy. War has become the result, and a cause, of the limitations placed on the forethinker. Power and its assistant, force, the natural enemies of intelligence, have become more serious as "the mental processes activated in the pursuit and consolidating of power are essentially short range" (p. 99). But it will not do to join the great chorus of those who create a crisis by saying there is a crisis.

It remains for me to thank again those who assisted in publishing the articles which have been a basis of this volume and in particular to thank those of my family and others who gave their assistance during a period of prolonged illness.

H. A. I.

Note

This volume was just about to go to press when Dean Innis died. In spite of great difficulty he had worked on it through the summer of 1952, and had finished correcting the proofs. These essays are the last work seen through the press by this distinguished Canadian scholar, whose researches were cut short by his untimely death.

CHAPTER ONE

The Strategy of Culture

With Special Reference to Canadian Literature—A Footnote to the Massey Report

Pay them well; where there is a Maecenas there will be a Horace and a Virgil also.

—Martial

Complaints are made that we have no literature; this is the fault of the Minister of the Interior.

—Napoleon

The title of this chapter may be regarded as an illustration of the remark of Julien Benda concerning "the *intellectual organization of political hatreds*"[1] and as a further effort to exploit Canadian nationalism. "Political passions rendered universal, coherent, homogeneous, permanent, preponderant—everyone can recognize there to a great extent the work of the cheap daily political newspaper."[2] Whistler[3] and others have contended that art is not to be induced by artificial tactics. They have pointed to Switzerland as a country without art and it has interesting parallels with Canada, a country of more than one language, a federation, and dependent on the tourist trade. A distinguished Canadian painter has remarked: "I am not sure that future opinion of the contemporary art of our day will not consider the advertising poster, the window and counter card as most representative."[4]

Printers' ink threatens to submerge even the literary arts in Canada and it may seem futile to raise the question of cultural possibilities. The power of nationalism, parochialism, bigotry, and industrialism may seem too great.

Cheap supplies of paper produce pulp and paper schools of writing, and literature is provided in series, sold by subscription, and used as an article of furniture. Almost alone Stephen Leacock, by virtue of his mastery of language, escaped into artistic freedom and was recognized universally and even he, as Peter McArthur pointed out, never attacked a publisher.

But we can at least point to the conditions which seem fatal to cultural interests. We can appraise the cultural level of the United States and appreciate the importance of New York as a centre for the publication of books and periodicals, the effects of the higher costs of commercial printing in Chicago, and the dangers to literature and the drama of reliance on the authoritative finality of New York newspaper critics. We should be able to escape the influence of a western American news agency which advised that if you want it to sell "put a New York date line on it."

We can point to the dangers of exploitation through nationalism, our own and that of others. To be destructive under these circumstances is to be constructive. Not to be British or American but Canadian is not necessarily to be parochial. We must rely on our own efforts and we must remember that cultural strength comes from Europe.[5] We can point to our limitations in literature and to the consequent distortions incidental to the impact of mechanization, notably in photography. The story has been compelled to recognize the demands of the illustration and has become dominated by it.[6] The impact of the machine has been evident in the dependence of Edgar Wallace and Phillips Oppenheim and dictators of the quick action novel on the dictaphone.[7] An emphasis on speed and action essential to books produced for individual reading weakens the position of poetry and the drama particularly in new countries swamped by print.

Burckhardt[8] in his studies of Western civilization held that religion and the state were stable powers striving to maintain themselves and that civilized culture did not coincide with these two powers, that in its true nature it was actually opposed to them. "Artists, poets and philosophers have just two functions, i.e., to bring the inner significance of the period and the world to ideal vision and to transmit this as an imperishable record to posterity." In the words of Sir Douglas Copland, summarizing the philosophy of P. H. Roxby, "A cultural heritage is a more enduring foundation for national prestige than political power or commercial gain."[9] "It is the cultural approach of one nation to another, which in the long run is the best guarantee for real understanding and friendship and for good commercial and political relations. In the past, it has been, on the whole, sadly neglected, and especially as between western Europe and China" (Roxby).[10] It has been scarcely less neglected as between Canada and the United States. In the long list of volumes of "The Relations of Canada and the United States" series, little interest is shown in cultural relations and the omission is ominous.

Inter-relations between American and Canadian publishing in the nineteenth century had significant implications for Canadian literature in the present century. In the nineteenth century the tyranny of the novel in England had been built up in part because of inadequate protection to English playwrights from translations of French plays, production of which had been systematically encouraged in France,[11] and by a monopoly of circulating libraries protected by the high price of the three-volume novel which made it, therefore, cheaper to rent than to buy books.[12] Restrictive effects of high prices on exports of books from Great Britain, absence of circulating libraries in the United States, lack of protection to foreign, especially English books before the enactment of copyright legislation in America in 1891, and section 5 of the American Copyright Act, May 31, 1790, which was "an invitation to reprint the work of English authors," were factors responsible for large-scale reprinting of English works in the United States and for the publication of English works first in the United States.[13]

In 1874 legislation in the United States reduced postage on newspapers issued weekly or oftener to two cents a pound without regard to the distance carried. Under an act of March 3, 1879 (par. 14), second-class mail matter "must be regularly issued at stated intervals as frequently as four times a year, and bear a date of issue, and be numbered consecutively." Again, on July 1, 1885, postal charges on paper-covered books were reduced from two cents per pound to one cent and cloth-bound books were carried at eight cents per pound. The legislation reflected the demands of a vigorous cheap book publishing period, concentrating on English or foreign books for which a market had been created by established publishers.

In the ultimate development of the publication of English books previous to the Copyright Act in 1891, Canadians, emigrants to the United States and undisciplined by the demands of its distributing machinery, played an important role. George Munro, a mathematics teacher in the Free Church College, Halifax, who had emigrated to New York and acquired experience in the handling of dime novels in the firm of Beadle and Adams and in the publishing of the *Fireside Companion*, a family newspaper started in 1867, launched the "Seaside Library," a quarto, two or three columns to the page with cheap paper, on May 28, 1877. It was estimated that 645 pages in a regular edition could be printed in 152 pages quarto. As a result of saturation of the market for quartos in the latter part of 1883, Munro started a pocket-size edition in spite of the higher costs of manufacturing. In 1887 he cut wholesale prices from twenty and twenty-five cents to ten cents and from ten cents to five cents, and in 1889 sought protection by publishing a monthly "Library of American Authors," cheap cloth-bound twelvemos, "sold by the ton." In 1890 Munro sold the "Seaside Library" to J. W. Lovell,[14] on a three-year option to repurchase arrangement, for $50,000 plus $4,500 monthly. It was estimated that, by 1890, 30,000,000 volumes of the "Seaside Library" had been sold, chiefly through the American News Company.

J. W. Lovell was the son of John Lovell, who in 1872 had a printing shop on the American side near Montreal at which he printed British copyright works, free of copyright, and imported them into Canada under 12.5 percent duty to be sold at a lower price than editions imported from Great Britain.[15] The son moved to New York in 1875 and engaged in the sale of cheap unauthorized editions. After a failure in 1881 he followed the German plan of producing cheap handy books with neat covers and, in 1882, started publication of handy twelvemos in "Lovell's Library," paper-covered books selling at twenty cents, and "Lovell's Standard Library," cloth-bound at one dollar. In 1885 he concentrated on "Lovell's Library" and sold the remainder of his business to Belford and Clarke. This became a most popular series selling about seven million volumes annually. As a result of the reduction of prices by George Munro in 1887 competition became more intense and in 1888 Lovell bought the "Munro Library"[16] from Norman W. Munro, the brother of George Munro. The "Munro Library" in pocket-size books had been started in 1884 when the owner had returned to the business after failing with the "Riverside Library," sold between 1877 and 1879. With control over the "Seaside Library" acquired in 1890, and over the plates and stock of other cheap book publishers by purchase or rental to the extent of over half the titles of cloth-bound books and over three-fourths of the titles of paper-covered books, and supported by the Trow Printing Company, Lovell organized the United States Book Company with a reported capital of $3,500,000.

Alexander Belford and James Clarke, members of a firm of Belford brothers in Toronto, moved to Chicago and organized Rose, Belford and Company; it was reorganized in 1879 after a failure as Belford Clarke and Company. They became publishers of "railroad literature" and built up an elaborate retail system developing a policy of selling to the book trade at artificially high prices, first to jobbers, and then to the regular trade, and later at extremely low prices through dry-goods and department stores. Showy bindings contrasted with the woodpulp, clay, and straw paper inside the books. In 1885 they acquired "Lovell's Standard Library" and became the largest producers of cheap cloth-bound twelvemos. As a result of the intensive price cutting after 1887 they failed in 1889.

In the absence of copyright on foreign books, publishers were compelled to rely on their only means of protection, namely, cheapness based on mass production. With efficient systems of distribution through the American News Company and the post office, equipment was steadily improved; cylinder presses were first installed in 1882 and in 1886 three cheap library publishers had their own typesetting, printing, and binding plants. The cheapest variety of paper was used and slight attention was given to proof reading and corrections. Paper manufacturers were compelled to sell their fine book papers chiefly to the large printing houses and the periodical publishers. Stereotype establishments or "sawmills" began to sell plates to publishers who then issued their own editions. Typographical unions[17] complained, and, following the sharp reduction in prices, recognized the importance of copyright. With lower postal rates on paper-covered editions, and prices from one-sixth to one-tenth those of cloth-bound volumes, it was estimated that almost two-thirds of a total of 1,022 books published in 1887 were issued in the cheap libraries. Demands for new titles led to the publication of poorer classes of fiction.[18] The technological changes which lowered the prices of paper[19] and of printing widened the gap between the supply of written material and the demand of readers and intensified the need for non-copyright foreign books. Yet the supply of foreign material was limited, the market for lower grade fiction was saturated, it was no longer possible to increase sales by changing formats from quarto to twelvemo, deterioration of paper was not sufficiently rapid, and finally newspapers expanded to absorb supplies of newsprint. Publishers were now compelled to emphasize American writers, to whom copyright was paid. The basis was laid for the supremacy of the periodical, with significant consequences for American and Canadian literature. National advertising steadily advanced to impose its demands on the reading material of the periodical. The discrepancy between prices of books in England and in the United States gradually lessened. The three-volume novel disappeared in England as prices were levelled with those in the United States after the Copyright Act of 1891. To secure copyright it was necessary to print books in the United States.[20]

In the last decade of the nineteenth century the advantages of cheap newsprint, of cheap composition following the invention of the linotype, and of the fast press as the basis of large circulations were being fully exploited by newspapers. Every conceivable device to increase circulation was pressed into service, notably in the newspaper war between Pulitzer and Hearst in the late nineties in New York City, including sensational headlines, the comics, and the Spanish American War. Crusades were started in every direction to enhance goodwill for newspapers.

The sudden improvement in technology in the production of newspapers was accompanied by an increase in magazine readers. The weekly was replaced by the monthly which became a leading factor in modern publishing. The Copyright Act of 1891, in itself a recognition of the problem of creating a supply of American writers,[21] was followed by the training of an army of fiction writers who by 1900 met the demands of magazines. Muck-raking magazines[22] were supported by experienced newspaper men such as Lincoln Steffens (who wrote a series on "The Shame of the Cities"). They followed the tactics, particularly of the Hearst newspapers, in the struggle for circulation.[23] McClure, for instance, applied the sensational methods of the cheap newspaper to the cheap and new magazine. He sponsored a reform wave which was effectively exploited by Theodore Roosevelt. He built up circulation by paying enormous sums to famous writers and trying to corner a market in them. As a former peddler of coffee pots, he knew the demands of people on farms and in small towns.[24] Munsey,[25] in the all-fiction magazine which followed the Sunday magazine section of the newspaper with smooth paper and clearer half-tones, made fiction the basis of circulation and earning power by 1896.[26]

The position of women as purchasers of goods led to concentration on women's magazines and on advertising. In Philadelphia, Curtis developed the great discovery—reading matter trailed through a periodical compelled readers to turn the pages and to look at the advertising which made up most of the page—into an extensive magazine business.[27] Through the national magazine,[28] advertisers such as the manufacturers of pianos, high cost two-wheeled bicycles, and other commodities were able to reach a large market at less cost than through the daily newspaper and to concentrate on more attractive layouts appealing to people in higher income brackets. The national magazine made a systematic attack on older advertising media. Religious papers dependent on patent medicine advertising felt the effects of a crusade of the *Ladies' Home Journal*, which in 1892[29] refused to handle medical advertising and exposed widely advertised preparations by printing chemical analyses. With the growth of large-scale printing, the printer assumed the direction of advertising and displaced the single advertiser and agency. Specialization of printing and increased pressure of overhead costs necessitated effective control of publications. Lorimer, an able writer of advertisements, became editor of the *Saturday Evening Post* and gave advertisements the personality of articles.[30] A four-colour printing press costing $800,000 and a new building in 1910 led the Curtis publications to add a third magazine to cover agriculture.[31]

The average circulations of magazines increased from 500,000 to 1,400,000 in the period from 1905 to 1915 and following the boom beginning in 1922 reached 3,000,000 by 1937.³² The *Reader's Digest* was started in 1922, *Time* in 1923, and the *New Yorker* in 1925. Extension of education and increased use of text-books conditioned youth to acceptance of the printed word and to magazine consumption. The demand for writers exceeded the supply. After the First World War, women's magazines, which had begun as pattern makers in the *Delineator* and other Butterick papers, gained conspicuously in circulation. Women's magazines reached the largest circulations, paid most highly for articles, and were the chief market for writers. Competition between magazines for writers with an established reputation brought sky-rocket prices.³³ The sale of film rights to popular novels brought even more than that of serial rights. An average bestseller in "the slicks" with serial rights, movie, book, and other rights brought returns varying between $70,000 and $125,000. Writers concentrated on magazines rather than books.³⁴

Writing for the great popular magazines built up on advertising implied assiduous attention to their requirements on the part of writers and editors. Dullness was absolutely abhorrent. Serial installments involved consideration of appropriate terminal points at which intense interest might be sustained for the next number. Magazines with the largest circulation were able to carry longer fiction by writers with an established reputation but tended to reduce installments and stories from 12,000 to 5,000 or 4,500 words.³⁵ Since dependence on advertising meant that the magazine "expands and contracts with the activity of the factory chimney"³⁶ writers were particularly affected by fluctuations of the business cycle. The reputations of authors were built up through advertising by editors of magazines who were thus enabled to sell advertising material, and stories³⁷ became commercialistic. George Ade could write "I guess I can now sell anything I write, even if it's good."³⁸

The influence of the newspaper and advertising on the magazine was developed to a sophisticated level in the twenties when magazines such as the *New Yorker* playfully exposed the foibles of its advertisers and advertisers exploited the foibles of the magazine. More recently the campaign of the *New Yorker* against loud speaker advertising in public buildings has not been unrelated to competition for advertising—all of course in the spirit of good clean fun. The rigid limitations in style of advertising copy enabled the *New Yorker* to succeed by emphasizing the independence of the editor from the business office, and by developing a new style of writing which in turn led to a revolution in the style of advertising copy. In the *Smart Set* and the *American Mercury* H. L. Mencken, a Baltimore newspaperman, was successful in building up circulation in a direct attack on the limitations of a society dependent on advertising. In reviewing books for newspapers he had become familiar with trends in literature and he attracted to the *Smart Set* new authors unable to secure publication with old firms and willing to acquire prestige in lieu of high rates of pay. As a columnist Mencken had also gained an intimate knowledge of libel laws. Of German descent, he had suffered from the frenzied propaganda of the First World War. The *American Mercury* was started in 1924 as a fifty-cent magazine and practically doubled its average monthly circulation from 38,694 to 77,921 by 1926.[39] Debunking became a new word and a profitable activity. In developing the *American Mercury* as a quality magazine designed to make the common man respectable,[40] Mencken pursued his attacks on the puritanical and on the English book to the point of recognizing in a powerful fashion the new language of the newspaper and the magazine in his *American Language*.

The women's magazines began to feel the restraining influence of puritanism and its effects on advertising. Bok became concerned with the importance of sex education. Theodore Dreiser, editor of *Delineator*, came into conflict with censorship regulations in his novels and triumphantly conquered in *An American Tragedy*. Mencken, in the tradition of Mark Twain and Ambrose Bierce, secured the support of the Authors' League for Dreiser's position.[41] The Calvinistic obsession of hypocritical people with the subject of sex[42] became the centre of attack by Dreiser as chief artist and Mencken as high priest, determined to defeat "the iron madonna who strangles in her fond embrace the American novelist" (H. H. Boyesen). With a shrewd appreciation of the advertising value of censorship regulations Mencken seized upon the occasion of the banning of a copy of the *American Mercury* to attack the Boston Watch and Ward Society as the stronghold of Catholic and Protestant puritanism.[43] His active interest in the Scopes trial, following a law enacted in Tennessee on March 21, 1925, against the teaching of evolution was a part of the general strategy against religious bigotry.

Decline of the practice of reading aloud led to a decline in the importance of censorship. The individual was taken over by the printing industry and his interest developed in material not suited to general conversation. George Moore in England and H. L. Mencken in the United States exploited the change in their attacks on censorship. Censorship could no longer be relied upon to secure publicity. Significantly the advertiser had contributed to a change of atmosphere and women no longer feared to smoke cigarettes in public.

Even before the Copyright Act, the effects of advertising, as reflected in the newspaper and the magazine, on the writer had important implications for the book. "Most people now do not read books, but read magazines and newspapers" (H. C. Baird).[44] Limited distributing facilities for books evident in the high costs of book agents and subscription publishing[45] in the nineties, and the development of special publishers of text-books in the early part of the century were gradually being offset by department stores. Small retail stores for books could not compete with rents paid by diamonds, furs, and bonds. Mail order business in books expanded in the early 1900s but the results were perhaps evident in the remark of a publisher's reader, "this novel is bad enough to succeed."[46] W. D. Howells wrote in 1902: "Most of the best literature now sees the light in the magazines, and most of the second best appears first in book form." The increasing importance of apartment buildings and lack of space for shelves supported the rapid development of the lending library in the twenties. Book clubs increased rapidly[47] after 1926 as a means of securing the economies of mass production. Nevertheless, the inadequacy of book distributing machinery and dependence on British and Continental devices[48] showed the limitations of the book in contrast with the newspaper and the magazine. Publishing firms such as Doubleday, Page and Company entered on policies of direct vigorous advertising, which built up, for instance, the success of O. Henry,[49] but their most significant results were in less obvious directions.

The experience of the prominent publishing firm of Scribner's illustrates directly the impact of advertising on the newspaper and the magazine and in turn on the book. Roger Burlingame,[50] trained in a newspaper office, and M. E. Perkins, a reporter on the *New York Times*, exercised a powerful influence on publications of the firm. Perkins was concerned to arouse a consciousness of the value and importance of the native note in opposition to the imitation of English and European models and "the cynical disparagement of American materialism."[51] To him great books were those which appealed to both the literati and the masses. The book-buying public was made up of fairly successful people but to Perkins the reading of Thomas Wolfe's books "to pieces" in the libraries reflected the truer sense of life of people in the lower economic level.[52] While he condemned the mad pursuit of best-sellers which developed during the boom period of the twenties and the newspaper policy of playing up the work of authors of best-sellers and criticized the Book of the Month Club of concentrating the attention of the public on one book a month,[53] he was concerned primarily with the newspaper public. Writers from the newspaper field included Hemingway, Edmund Wilson, Stanley Pennell, Stephen Crane, and Dreiser. It was his opinion that the teaching of literature and writing in the colleges compelled students to see things through a film of past literature and not with their own eyes. Two years with a newspaper were better than two years in college.[54] He favoured what Irving Babbitt called "art without selection." The demands of commercialism were evident more directly in the avoidance of controversy. "The sales department always want a novel.

They want to turn everything into a novel."[55] The public and the trade preferred books of 100,000 words and works of 25,000 to 30,000 words were padded to give the appearance of books of a larger size.

An orderly revolt against commercialism was significantly delayed and frustrated in literature possibly more than in any other art. Henry James had escaped to England and in the period after the First World War Ezra Pound and T. S. Eliot followed. "The historians of Wolfe's era . . . all record this strange phase of our cultural adolescence; the same sad and distraught search for foreign roots."[56] "You could always come back" (Hemingway). But in the words of Pound: "We want a better grade of work than present systems of publishing are willing to pay for."[57] "The problem is *how*, how in hell to exist without over-production."[58] "The book-trade, accursed of god, man and nature, makes no provision for *any* publication that is not one of a series. . . ."[59] "The American law as it stands or stood is all for the publisher and the printer and all against the author, and more and more against him just in such proportion as he is before or against his time."[60] Books by living authors were, he claimed, kept out of the United States and "the tariff, which is iniquitous and stupid in principle, is made an excuse."[61] Even in Great Britain from about 1912 to 1932 booksellers did "their utmost to keep anything worth reading out of print and out of ordinary distribution." "Four old bigots" of Fleet Street practically controlled the distribution of printed matter in England.[62] Criticism was related to publishers' advertising.[63]

The distorting effects of industrialism and advertising on culture in the United States have been evident on every hand. Architecture as a sort of tyrant of the arts had the advantage of the utilitarian demands of commerce. Painting and sculpture as allied to it had the support of collectors, private and public, and the encouragement of awards and prizes.[64] Poetry was the subject of paragraphers' jokes, a space filler for magazines[65] and "must appeal to the barber's wife of the Middle West."[66] "Poetry had no one to speak for it."[67] In the drama the lack of interest of actors in modern art[68] and the support of tradition involved effective reliance on Shakespeare and a terrific handicap to playwrights.[69] The commercial theatre manager and the newspaper critic have been reluctant to recognize the vitality of a demand for the imaginative artistic work of the little theatre[70] particularly in competition with the cinema. In the words of George Jean Nathan the talking picture may be "the drama of a machine age designed for the consumption of robots" and the theatre may have gained enormously by the withdrawal of "shallow and imbecile audiences," but the change has been costly and painful.[71]

The overwhelming pressure of mechanization evident in the newspaper and the magazine has led to the creation of vast monopolies of communication. Their entrenched positions involve a continuous, systematic, ruthless destruction of elements of permanence essential to cultural activity. The emphasis on change is the only permanent characteristic. Thomas Hardy complained that narrative and verse were losing organic form and symmetry, the force of reserve, and the emphasis on understatement, and becoming structureless and conglomerate.[72]

The guarantee of freedom of the press under the Bill of Rights in the United States and its encouragement by postal regulations has meant an unrestricted operation of commercial forces and an impact of technology on communication tempered only by commercialism itself.[73] Vast monopolies of communication have shown their power in securing a removal of tariffs on imports of pulp and paper from Canada though their full influence has been checked by provincial governments especially through control over pulpwood cut on Crown lands. The finished product in the form of advertisements and reading material is imported into Canada with a lack of restraint from the federal government which reflects American influence in an adherence to the principle of freedom of the press and its encouragement of monopoly. Sporadic attempts have been made to check this influence in Canada as in the case of the banning of the Hearst papers in the First World War and in the imposition by the Bennett administration of a tariff based on advertising content in American periodicals. Protests are made by institutions against specific articles in American periodicals but without significant results other than that of advertising the periodical. To offset possible handicaps Canadian editions of *Time*, *Reader's Digest* and the like are published. Canadians are persistently bombarded with subscription blanks soliciting subscriptions to American magazines, and their conversation shifts with regularity following the appearance of new jokes in American periodicals. Canadian publications supported by the advertising of products of American branch plants and forced to compete with American publications imitate them in format, style and content. Canadian writers must adapt themselves to American standards.[74] Our poets and painters are reduced to the status of sandwich men. The ludicrous character of the problem may be shown by stating that the only effective means of sponsoring Canadian literature involves a rigid prohibition against all American periodicals with any written material and free admission to all periodicals with advertising only. In this way trade might be fostered and Canadian writers left free to work out their own solutions to the problems of Canadian literature. Indeed they would have the advantage of having access to the highly skilled examples of advertiser's copy.

Publishers' lists in Canada are revealing in showing the position of American branches of American agencies in the publication of books. Advertising rates for a wide range of commodities, determined by newspapers and magazines particularly in relation to circulation, are such as to make it extremely difficult for publishers to compete for advertising space, particularly as book advertising is largely deprived of the powerful force of repetition.[75] Moreover, the demands of a wide range of industries for advertising compete directly and effectively for raw materials, paper, capital, and labour entering into the production of books, and restrict the possibility of advertising them. American devices such as book clubs and the mass production of pocket books to be sold on news-stand and in cigar stores and drug stores have immediate repercussions in Canada. The extreme importance of book titles—perhaps the most vital element in American literature—evident in the changing of titles of English books in the United States and of American books in Great Britain and in the interest of the movie industry in the publishing field,[76] is felt in Canada also. In the field of the newspaper, dependence on the Associated Press and other agencies, on the *New York Times*,[77] and other media needs no elaboration. In radio and in television accessibility to American stations means a constant bombardment of Canadians.

The impact of commercialism from the United States has been enormously accentuated by war. Prior to the First World War the development of advertising[78] stimulated the establishment of schools of commerce and the production of text-books on the psychology of advertising. European countries were influenced by the effectiveness of American propaganda. Young Germans were placed with American newspaper chains and advertising and publishing agencies to learn the art of making and slanting news. American treatises on advertising and publicity were imported and translated. American graduate students were attracted to Germany by scholarships and experiments in municipal government. In turn, German exchange professorships were established, especially with South American universities. The Hamburg-American Lines became an effective propagandist organization. But German experience[79] proved much too short in contrast with that of American[80] and English propagandists,[81] though their effectiveness is difficult to appraise since the estimates have been provided chiefly by those responsible for the propaganda.

American propaganda[82] after the First World War became more intense in the domestic field. Its effectiveness was evident in the emergence of organizations representing industry, labour, agriculture, and other groups. The Anti-Saloon League pressed its activities to success in prohibition legislation. In the depression the American government[83] learned much of the art of propaganda from business and exploited new technological devices such as the radio. With the entry of the United States into the Second World War instruments of propaganda[84] were enormously extended.

The effects of these developments on Canadian culture have been disastrous. Indeed they threaten Canadian national life. The cultural life of English-speaking Canadians subjected to constant hammering from American commercialism is increasingly separated from the cultural life of French-speaking Canadians. American influence on the latter is checked by the barrier of the French language but is much less hampered by visual media. In the period from 1915 to 1920 the theatre in French Canada was replaced by the movie or French influence by American. With the development of the radio, protection of language enabled French Canadians to take an active part in the preparation of script and in the presentation of plays. During the Second World War the revue and the French-Canadian novel received fresh stimulus. The effects of American technological change on Canadian cultural life have been finally evident in the numerous suggestions of American periodicals that Canada should join the United States. It should be said that this would result in greater consideration of Canadian sentiment by American periodicals than is at present the case when it probably counts for less than that of a religious sect.

The dangers to national existence warrant an energetic programme to offset them. In the new technological developments Canadians can escape American influence in communication media other than those affected by appeals to the "freedom of the press." The Canadian Press has emphasized Canadian news but American influence is powerful.[85] In the radio, on the other hand, the Canadian government in the Canadian Broadcasting Corporation has undertaken an active role in offsetting the influence of American broadcasters. It may be hoped that its role will be even more active in television. The Film Board has been set up and designed to weaken the pressure of American films. The appointment and the report of the Royal Commission on National Development in the Arts and Sciences imply a determination to strengthen our position. The reluctance of American branch plants to support research in Canadian educational institutions has been met by taxation and federal grants to universities. Universities have taken a zealous interest in Canadian literature but a far greater interest is needed in the whole field of the fine arts. Organizations such as the Canadian Authors' Association have attempted to sponsor Canadian literature by the use of medals and other devices. The resentment of English and French Canadians over the treatment of a French-Canadian play on Broadway points to powerful latent support for Canadian cultural activity.

We are indeed fighting for our lives. The pernicious influence of American advertising reflected especially in the periodical press and the powerful persistent impact of commercialism have been evident in all the ramifications of

Canadian life. The jackals of communication systems are constantly on the alert to destroy every vestige of sentiment toward Great Britain holding it of no advantage if it threatens the omnipotence of American commercialism. This is to strike at the heart of cultural life in Canada. The pride taken in improving our status in the British Commonwealth of Nations has made it difficult for us to realize that our status on the North American continent is on the verge of disappearing. Continentalism assisted in the achievement of autonomy and has consequently become more dangerous. We can only survive by taking persistent action at strategic points against American imperialism in all its attractive guises. By attempting constructive efforts to explore the cultural possibilities of various media[86] of communication and to develop them along lines free from commercialism, Canadians might make a contribution to the cultural life of the United States by releasing it from dependence on the sale of tobacco and other commodities which would in some way compensate for the damage it did before the enactment of the American Copyright Act.

Notes

1. Julien Benda, *The Great Betrayal* (London, 1928), p. 21.
2. Ibid., p. 7.
3. J. M. Whistler, *The Gentle Art of Making Enemies* (New York, 1904).
4. William Colgate, *C. W. Jeffreys* (Toronto, n.d.), p. 28.
5. "Until the English visitor to America comprehends that he is in the midst of a civilization totally different from anything he has known on our side of the Atlantic, he is exposed to countless shocks" (Sir John Pollock, Bt., *Time's Chariot* [London, 1950], pp. 184–85). Sir John regards the great difference as having developed since 1880 as a result of the Civil War and foreign immigration. In England, with a background of feudalism, it seems possible to keep political differences and personal relationships in separate departments.
6. Whistler's complaint that painting was subordinate to literature must be offset by the account of Newman Flower of Cassell & Co. He resorted to a *cliché* department or "bank" of illustrations built up since 1870, selected a promising illustration, and asked a young writer to write around it. *Just As It Happened* (London, 1951), p. 27.
7. Ibid., p. 40. On the other hand Edgar Wallace protested that dictaphone stuff was "good Wallace publicity. I write my best stuff with a pen." Reginald Pound, *Their Moods and Mine* (London, 1939), p. 233. "Dictation always is rubbish" (George Moore). Ibid., p. 112. As a result of the influence of the newspaper on reading, novels have been written to be read rapidly and consequently emphasize length and description. "I do not want literature in a newspaper" (E. L. Godkin).
8. See Jacob Burckhardt, *Force and Freedom: Reflections on History* (New York, 1943).

The Strategy of Culture 15

9. D. B. Copland, "Culture versus Power in International Relations" in *Liberty and Learning: Essays in Honour of Sir James Hight* (Christchurch, 1950), p. 155.

10. Ibid., p. 154.

11. In France the Théâtre Français was subsidized by the government, and the Society of Dramatic Authors founded by Beaumarchais and reorganized by Scribe in the nineteenth century fostered an interest in plays rather than novels. See Brander Matthews, *Gateways to Literature and Other Essays* (New York, 1912), p. 41 and also H. A. Innis, *Political Economy in the Modern State* (Toronto, 1946), pp. 35–55.

12. See introduction by Graham Pollard to I. R. Brussel, *Anglo-American First Editions, 1826–1900* (New York, 1935), p. 10.

13. Ibid., p. 11. See also H. A. Innis, *The Bias of Communication* (Toronto, 1951), pp. 171–72.

14. The funds became the basis of a substantial gift to Dalhousie University.

15. R. H. Shove, *Cheap Book Production in the United States, 1870 to 1891* (Urbana, 1937), p. 75. This book is a mine of information.

16. This included 855 sets of plates and 1,500,000 copies of books for which $250,000 was paid.

17. The unions were at first opposed to the Copyright Act but became active in its support; see G. A. Tracy, *History of the Typographical Union* (Indianapolis, 1913), p. 450.

18. Brussel, *Anglo-American First Editions*, p. 19.

19. In 1871, newsprint straw paper was twelve cents per pound, fine book paper sixteen to seventeen cents; in 1875 newsprint was nine cents, machine-finish book paper ten to eleven cents; in 1889 newsprint was three and one-quarter cents and calendared book paper six and one-half to seven and one-half cents (Shove, *Cheap Book Production*, p. 4).

20. Cheap unauthorized editions disappeared and the works of authors such as Kipling, which had sold widely in pirated editions, were sold at higher prices and in smaller numbers.

21. The suit brought against the *New York World* by Harriet Monroe for printing her ode presented at the opening of the Chicago World's Fair and the award of $5,000 damages strengthened the position of authors (Harriet Monroe, *A Poet's Life: Seventy Years in a Changing World* [New York, 1938], pp. 139–43).

22. C. C. Regier, *The Era of the Muckrakers* (Chapel Hill, 1932).

23. H. L. Mencken, *Prejudices, First Series* (New York, 1929), p. 175.

24. S. S. McClure, *My Autobiography* (New York, 1914).

25. F. A. Munsey, *The Founding of the Munsey Publishing House* (New York, 1907); also George Britt, *Forty Years—Forty Millions: The Career of Frank A. Munsey* (New York, 1935).

26. Algernon Tassin, *The Magazine in America* (New York, 1916), pp. 342–43.

27. Arthur Train, *My Day in Court* (New York, 1929), p. 419.

28. Frank Presbrey, *The History and Development of Advertising* (New York, 1929), p. 339.

29. Ibid., pp. 531–32. See *The Americanization of Edward Bok: The Autobiography of a Dutch Boy Fifty Years After* (New York, 1937). Also Edward W. Bok, *A Man from Maine* (New York, 1923). The campaign against patent medicines provoked the announcement by Eugene Field of the engagement of the granddaughter of Lydia W. Pinkham to Edward W. Bok, the editor of the *Ladies' Home Journal*.

30. Bok, *A Man from Maine*, p. 171. "The secrets of success as an editor were easily learned; the highest was that of getting advertisements. Ten pages of advertising made an editor a success; five marked him as a failure" (Henry Adams, *The Education of Henry Adams: An Autobiography* [Boston, 1918], p. 308). "The art of advertising has outgrown the art of creative writing. . . . Three-fourths of the income of the magazines comes from their advertisers. . . . just take the advertising and rewrite it" (W. E. Woodward, *Bunk* [New York, 1923], p. 51).

31. Bok, *A Man from Maine*, p. 183.

32. Train, *My Day in Court*, p. 421.

33. Fairfax Downey, *Richard Harding Davis, His Day* (New York, 1933), p. 219.

34. *Ibid.*, pp. 430–31, 433.

35. Train, *My Day in Court*, pp. 423–25. In England Gilbert Frankau held that the serial market was disappearing because readers of monthly magazines would not wait and newspapers preferred the short story "in these days of so much frontpage excitement" (Pound, *Their Moods and Mine*, p. 241).

36. Train, *My Day in Court*, p. 420. The limited circulation of Canadian magazines makes for a seasonal expansion. Advertising is sufficient only during the period of the two or three months before Christmas to warrant a full-fledged interest in features, especially short features. Longer features appear after the holiday season.

37. Train, *My Day in Court*, p. 440.

38. F. W. Wile, *News Is Where You Find It* (Indianapolis, 1939), p. 36.

39. W. Manchester, *Disturber of the Peace: The Life of H. L. Mencken* (New York, 1951), p. 15.

40. Ibid., p. 155.

41. Ibid., pp. 93–94.

42. Ibid., p. 101.

43. Ibid., p. 207. See an account of the failure of attempts by Covici, Friede to secure suppression of Radclyffe Hall's *Well of Loneliness* by the Boston Watch and Ward Society (Donald Friede, *The Mechanical Angel* [New York, 1948], p. 94).

44. Cited in J. C. Derby, *Fifty Years among Authors, Books and Publishers* (New York, 1884), p. 559.

45. Subscription selling was accompanied by a development of techniques of salesmanship and depended for its success to an important extent on snob appeal, particularly the prestige attached to owning a large book among the relatively illiterate. Estes and Lauriat of Boston, prominent subscription book agents, who came under the control of Walter Jackson and Harry E. Hooper after 1900, were active in developing schemes for the sale of the *Encyclopaedia Britannica* in connection with the London *Times*.

The Strategy of Culture   17

46. W. H. Page, *A Publisher's Confession* (New York, 1905), p. 27.
47. E. H. Dodd, *The First Hundred Years: A History of the House of Dodd, Mead, 1839–1939* (New York, 1939), p. 36.
48. O. M. Sayer, *Revolt in the Arts* (New York, 1930).
49. Train, *My Day in Court*, p. 439.
50. Roger Burlingame, *Of Making Many Books* (New York, 1946), p. 221.
51. J. H. Wheelock, *Editor to Author: The Letters of Maxwell E. Perkins* (New York, 1950), p. 8.
52. Ibid., p. 184.
53. Ibid., p. 128.
54. Ibid., p. 267. "What the eighteenth century thought simply vulgar, and the nineteenth gathered data from, has now become literary material; even the annals of the poor are to be short and simple no longer" (H. W. Boynton, *Journalism and Literature and Other Essays* [Boston, 1904], p. 164).
55. Wheelock, *Editor to Author*, p. 84.
56. Maxwell Geismar, *Writers in Crisis: The American Novel between Two Wars* (Boston, 1942), pp. 214 and *passim*.
57. *The Letters of Ezra Pound, 1970–1941*, ed. D. D. Paige (New York, 1950), p. 175.
58. Ibid., p. viii.
59. Ibid., p. 319.
60. Ibid., p. 52.
61. Ibid., p. 53. See J. L. May, *John Lane and the Nineties* (London, 1936), p. 159.
62. *Letters of Pound*, p. 239–40.
63. Ibid., p. 337.
64. Harriet Monroe, *A Poet's Life*, p. 241.
65. Ibid., p. 247. A study of the demands of space on Bliss Carman's poetry might prove rewarding.
66. Ibid., p. 288.
67. Ibid., p. 242.
68. Nathan refers to "the mean capacity of the overwhelming number of them, whatever their nationality. . . .the downright ignorance, often made so conspicuously manifest" (*The Intimate Note-books of George Jean Nathan* [New York, 1932], p. 144).
69. See a letter from Mrs. Fiske in Harriet Monroe, *A Poet's Life*, pp. 176–77.
70. Ibid., p. 419.
71. See St. John Ervine, *The Alleged Art of the Cinema* (n.p., March 15, 1934).

> Actors and actresses were certainly regarded with far greater interest than they are nowadays. The outstanding ones inspired something deeper than interest. It was with excitement, with wonder and with reverence, with something akin even to hysteria, that they were gazed upon. Some of the younger of you listeners would, no doubt, if they could, interrupt me at this point by asking, "But surely you don't mean, do you, that our parents and grandparents were affected by them as we are by cinema stars?" I would assure you that those idols were even more ardently worshipped than are yours. Yours after all, are but images of idols, mere shadows of glory. Those others were their own selves, creatures of flesh and

> blood, there, before our eyes. They were performing in our presence. And of our presence they were aware. Even we, in all our humility, acted as stimulants to them. The magnetism diffused by them across the footlights was in some degree our own doing. You, on the other hand, have nothing to do with the performances of which you witness the result. Those performances—or rather those innumerable rehearsals—took place in some far-away gaunt studio in Hollywood or elsewhere, months ago. Those moving shadows will be making identically the same movements at the next performance or rather at the next record; and in the inflexions of those voices enlarged and preserved for you there by machinery not one cadence will be altered. Thus the theatre has certain advantages over the cinema, and in virtue of them will continue to survive. (Sir Max Beerbohm in *The Listener* [Oct. 11, 1945], p. 397)

72. May, *John Lane and the Nineties*, p. 177.

73. See Upton Sinclair, *Money Writes! A Study of American Literature* (Long Beach, Calif., 1927).

74. One Canadian writer has complained of writing an article of 60,000 words for an American woman's magazine, cutting it to about 40,000 words to make two installments, and expanding it to 80,000 for the English market. Canadian writers should become efficient concertina players.

75. Wheelock, *Editor to Author*, p. 138.

76. See J. T. Farrell, *The Fate of Writing in America* (n.p., n.d.). also W. T. Miller, *The Book Industry* (New York, 1949).

> Before the war British publishers were often told by friends in the Canadian book trade that their public preferred the bigger, handsomer American book. They wanted value for money, and had been accustomed to measure value by size and weight. The story has often been told of the Canadian agent who handed one of his travellers an advance copy of a new book from a British publisher and asked, "How many can you sell of that?" The traveller, without opening the book, handed it back and said, "None." The agent, somewhat nettled, said, "None? But you haven't even looked at it." The traveller replied, "I don't need to. It doesn't weigh enough.'" (Michael Joseph, *The Adventure of Publishing* [London, 1949], p. 131)

77. It "set out to be dull and ponderous and it has achieved its purpose with a fidelity and thoroughness justly commanding the admiration of all lovers of bulk and solidity" (G. M. Fuller, "The Paralysis of the Press," *American Mercury*, Feb. 1926, p. 160).

78. Will Irwin, *Propaganda and the News* (New York, 1936). For an account of the influence of an advertising agent of a Canadian department store on advertising and journalistic ideas in England, see *Autobiography of a Journalist* edited with an introduction by Michael Joseph (London, n.d.), pp. 45, 50. The author, advised by the agent to begin journalism by writing advertisements for shopkeepers, used samples of full-page advertisements of the Canadian store (p. 66). Advertising methods were then introduced effectively in political campaigns.

79. G. S. Viereck, *Spreading Germs of Hate* (New York, 1930).

80. James R. Mock and Cedric Larson, *Words That Won the War: The Story of the Committee on Public Information, 1917–1919* (Princeton, N.J., 1939).

81. See Neville Lytton, *The Press and the General Staff* (London, 1921); Sir Campbell Stuart, *Secrets of Crewe House: The Story of a Famous Campaign* (London, 1920);

Walter Millis, *Road to War: America 1914–1917* (Boston, 1935); James Squires, *British Propaganda at Home and in the United States from 1914 to 1917* (Cambridge, Mass., 1935); H. D. Lasswell, *Propaganda Technique in the World War* (London, 1927).

82. See O. W. Riegel, *Mobilizing for Chaos: The Story of the New Propaganda* (New Haven, Conn., 1939).

83. See George Michael, *Handout* (New York, 1935); L. C. Rosten, *The Washington Correspondents* (New York, 1937).

84. See *Propaganda by Short Wave*, ed. H. L. Childs and J. R. Whitton (Princeton, N.J., 1943); C. J. Rolo, *Radio Goes to War: The "Fourth Front"* (New York, 1940).

85. "I am sceptical about the value of 90 per cent of press reports. Most of them tend to say enough to be misleading and not enough to be in any sense informative." Interview with a veteran Vancouver journalist. See M. L. Ernst, *The First Freedom* (New York, 1946) and Herbert Brucker, *Freedom of Information* (New York, 1949).

86. The problem to an important extent centers around the confusion as to the distinct possibilities of each medium. Literary agents deliberately exploit the demands of technological innovations, adapting the same artistic piece of work to the book, the magazine, and the film. See Curtis Brown, *Contacts* (London, 1935). Shaw refused to allow a play to be filmed stating that no one would go to see it after seeing it on the screen and that the author suffered because the play became dull with the dialogue left out (ibid., p. 51). The studios wanted "a big kick" at the end of every sequence of the film (ibid., p. 33). Mechanization demands uniformity. The newspapers are concerned with news and contemporary topics, and books, plays, films, and novels centre around newspaper owners. The book has been subordinated to the demands of advertising for the movies, business firms in centennial volumes, radio broadcasts, and articles from magazines. Bible scenes are exploited for plays and movies. Shakespeare's plays for actors are primarily studied in print as texts. Newspaper serials and radio scripts differ from novels and emphasize topics of the widest general interest. Any fresh idea is immediately pounced on and mauled to death. Irvin Cobb remarked concerning the dull conversation of Hollywood that the phrase coiners preserved silence until they had sold the wheeze themselves.

CHAPTER TWO

The Military Implications of the American Constitution

I

This chapter[1] is an attempt to understand the policies of the United States. In Canada we are under particular obligations to attempt such an understanding in our own interests as well as in the interests of the rest of the world. The difficulties involved in any country's understanding itself, particularly a country with a complex unstable history, are overwhelming and the most penetrating studies of the United States have been made by de Tocqueville, a Frenchman, and by Lord Bryce, an Englishman. A Canadian is too close to make an effective study but he has the most to gain from it. He is handicapped by tradition especially in English-speaking Canada, evident in the pervasive influence of those who left the United States after the American Revolution, namely the United Empire Loyalists, and by language in French-speaking Canada. The writer of this chapter can scarcely pretend to the necessary objectivity, nor, I suspect, can most of his readers. Nevertheless we must do our best.

Whatever our view about the American Revolution we must agree that it was achieved by a resort to arms against Great Britain. To the British it may have been a war of little consequence; we remember the remarks of an Englishman who when told that in the War of 1812 the British forces had burned Washington said he thought he had died in bed. To Americans the achievement was a result of desperate struggle. Revolutions leave unalterable scars and nations which have been burned over by them have exhibited the most chauvinistic brand of nationalism and crowd-patriotism.[2] These nations have developed highly depersonalized social relationships, political structures, and ideals and their counsels are determined most of all by spasms of crowd propaganda. "Public policy sits on the doorstep of every man's personal conscience. The citizen in us eats up the man."[3] The founders of the American Constitution appear to have recognized the danger by framing an instrument which put limits on the number of things concerning which a majority could encroach on the position of the individual.[4] But the extent of such protection has varied and declined with improvements in the technology of communication and the increasing powers of the executive, as Senator McCarthy has conspicuously shown.

Washington and his successors in the nineteenth century renounced an interest in Europe but steadily expanded their influence in the Americas following the increase in demand for new land on which to raise cotton. The demand implied steady expansion westward, in the south, and, in order to maintain a balance, in the north. In the south expansion was at the expense of the French empire, notably in Jefferson's administration when Louisiana was bought from Napoleon, and in the north at the expense of the British empire when Lewis and Clark were sent on a journey of exploration to the northwest and when John Jacob Astor established Astoria on the Columbia River. Later expansion in the south was safeguarded in the Monroe Doctrine, enunciated in 1823, which warned European powers to keep their hands off South America and was directed to the absorption of Texas, California, and other states at the expense of the Spanish empire and of Mexico. The remnants of a crumbling Spanish empire were finally taken over after the explosion of the *Maine* in Cuba ("Remember the *Maine*") and when Puerto Rico and the Philippines became American possessions. Expansion in the south to some extent intensified and to some extent eased the pressure on the British empire in the north. The line was eventually tightened to the present Canadian border and Alaska, "Seward's icebox," was purchased from Russia in 1867. These developments remind us of Disraeli's comment when Poland had been partitioned by European powers at a meeting at breakfast. "What will they have for lunch?"

II

The outbreak of the American Revolution marked a return to ideological warfare such as had largely disappeared in England after the Civil War.[5] Democratic nationalism and the mass army became the new basis of warfare.[6] George Washington, an officer in the British army in the Seven Years' War against the French, had gained experience which gave him the leadership of the Revolutionary Army. The immediate significance of the Revolution was evident in the position of this soldier from Virginia. A mass army could not be built up under a New England general.[7] As a result of success in arms he secured not only independence for the colonies but also a stable federal government. He presided over the Convention and was asked to take the chief position in the new government. An interest in western lands was not unrelated to his sympathy with the Federalists in their proposal for a strong central government with "powers competent to all general purposes," words included in a letter from him to Hamilton in 1783.[8] His sympathies found reflection in the views of delegates concerned about the dangers implicit in the radical character of state constitutions written by revolutionary legislatures. "Our chief danger rises from the democratic parts of our constitutions" (Edmund Randolph of Virginia to the Convention).[9] Conservatism and an emphasis on the theory of divided powers led to provisions strengthening the executive power, such as those making the President Commander in Chief of the Army and Navy and giving him control over patronage. The Secretaries of State and War were made responsible to the President alone and, with the exception of the Treasury Department, the precedent was followed in the establishment of new Cabinet posts. The President became a focus of executive power. The influence and character of Washington finally left their impression on the United States as he secured Virginia's acceptance of the Constitution in 1787 and gave leadership to the other states which followed.

In the work of establishing a nation, the influence and prestige of the first President left an indelible impression on the operation of government. However, Washington's efforts to secure the advice of the Senate as a sort of privy council were met with distrust. The decision of the Senate to receive reports of Cabinet ministers in writing and to exclude them from its meetings drove the Cabinet into the position of being the President's council. As a further guarantee against presidential interference, in Congress a system of committees was emphasized in which members were protected by secrecy from any group including the press.

John Adams, the second President (1797–1801), whose election implied a recognition of the role of New England in the Revolution and its aftermath, inherited the task of maintaining the prestige of the office, but he found it difficult to maintain the delicate balance between New England and the South, in the face of the power of Alexander Hamilton as a representative of industrial and commercial interests in the middle states. At Hamilton's insistence, Washington had agreed to call out the militia of four states to put down the Whisky Rebellion in 1794. In 1798 Hamilton advised his friends in the government to prepare for war with France, and Congress planned for a large emergency army and an increase in the regular army. Under his influence Washington agreed to head the army and by virtue of his prestige could insist on choosing his generals. Strife between Adams and Hamilton was followed by defeat of the former for a second term and by a weakening of the Federalist position.

In opposition to the centralizing tendencies of the Constitution, Jefferson (1801–1809) led a group whose views were reflected in the Declaration of Independence and the Articles of Confederation. He emphasized the position of the land, the small farmer, and the labourer against banking and the commercial interests. On his trip up the Hudson with Madison in 1791 he laid the foundations for the "longest-lived, the most incongruous, and the most effective political alliance in American history: the alliance of southern agrarians and northern city bosses."[10] In contrast with the Federalists who insisted that survival depended on the sword, Jefferson stated: "I hope no American will ever lose sight of the essential policy of interdicting in the seas and territories of both Americas, the ferocious and sanguinary contests of Europe." "Our first and fundamental maxim should be, never to entangle ourselves in the broils of Europe."[11] As a representative of the South, and in spite of his statement that "our peculiar security is in the possession of a written Constitution," he accepted the annexation of Louisiana and acquired the port of New Orleans without asking the question of constitutional propriety. To an alliance between the city bosses of New York and the South, he added the West.

After Jefferson's two terms, Madison, also a native of Virginia, became President (1809–1817) and acquired additional territory. On April 14, 1812, Congress formally divided West Florida at the Pearl River, annexing the western half to the new state of Louisiana, and, a month later, the eastern half to the Mississippi Territory. In 1813 the American army forced the Spanish garrison at Mobile to surrender and took possession. Henry Clay and the Committee on Foreign Affairs persuaded Congress to declare war on Great Britain on June 18, 1812. "The conquest of Canada is in your power." "This war, the measures which preceded it, and the mode of carrying it on, were all undeniably Southern and Western policy, and not the policy of the commercial states" (Josiah Quincy).[12] On December 5, 1814, Madison recommended liberal spending on the Army and the Navy and the establishment of military academies.

Following the two terms of Madison, Monroe, again a native of Virginia, and an officer in the Revolutionary Army, became President (1817–1825). The decline of the Federalist Party meant that there was no official opposition, and also no party discipline. The President was thus left without any device to secure cohesion in Congress. In the House of Representatives, for example, an Army bill, opposed by the President and the Secretary of War, was "carried notwithstanding many defects in the details of the bill by an overwhelming majority."[13] In 1822 Monroe recognized the independence of the Latin American republics which had been part of the Spanish empire, and, on the insistence of John Quincy Adams, included in his statement of the Monroe Doctrine on December 2, 1823, a protest against the encroachment of Russians in the northwest.

The success of the War of 1812 and the re-election of Monroe in 1820 finally destroyed the Federalist Party as a political factor. Decline in prestige and power of the congressional caucus opened the way for a free fight in 1824; New England influence was once more reflected in the election of John Quincy Adams, who like his father, John Adams, served only for one term (1825–1829).

His successor, Andrew Jackson (1829–1837), a native of South Carolina, had suffered at the hands of the British in the Revolutionary War. In the War of 1812 he had led western militiamen against the Indians of Georgia and Alabama and destroyed British troops under General Sir Edward Pakenham in New Orleans. In 1817 he pursued marauding Indians into Spanish territory, marched to Pensacola, and removed the Spanish governor. After his invasion of Florida he became military governor. As a national figure and a popular hero he introduced a system of military organization to national politics. Beginning in 1825 he built up a national political machine. A small, divided, virulent, and undisciplined[14] press which had contributed to the disappearance of the Federalist Party and a monopolistic Washington press were replaced by an organized party press designed to provide discipline and propaganda. The *National Intelligencer*,[15] the organ of Jefferson, Madison, Monroe, and J. Q. Adams, had been the oracle of war sentiment before and after 1812 and had a wide circulation for daily, semi-weekly, and weekly editions.[16] In opposition, Jackson and his followers established media to maintain a close contact with voters. After his election the *United States Telegraph* and the Washington *Globe* became administrative mouthpieces for partisan purposes.[17] Rewards were offered to strengthen the morale of the troops; "no plunder no pay." Political organizers in state politics such as Van Buren at Albany were brought to the national stage. In 1832 at the time of the nomination of Jackson for a second term, a system of nominating conventions was introduced in which a two-thirds rule was invoked to protect the position of the South. The news value of the system became evident in the emergence of the presidential candidate as the chief consideration of politics. Under Jackson and his successor, Van Buren (1837–1841), a representative of New

York State, campaign techniques were elaborated. Veto messages, written up by journalistic members of the Kitchen Cabinet for popular consumption, had a wide distribution. The difficulties of the system became evident when attempts were made to meet the demands of regional groups. The Tariff of Abominations, and the opposition to Vice President Calhoun of South Carolina in the nullification controversy, made the latter a defender of state rights and led to the enactment of the Force Act by which the President was given authority to call out the Army and Navy to enforce laws of Congress. The dragon's teeth of secession were sown.

To meet the type of organization built up in support of Jackson and Van Buren, an attempt was made to establish a Whig Party, based chiefly on anti-Masonic feeling,[18] following the contest of 1836. In New York State, Seward and Weed, to weaken the position of Van Buren and to exploit the news value of a war hero, secured the nomination of W. H. Harrison, who had been engaged in a battle with the Indians at Tippecanoe Creek in 1811, and was promoted to command the Army of the Northwest in the War of 1812. A vigorous campaign with an emphasis on such slogans as "log cabin and hard cider" led to his election in 1841 but his death shortly afterwards meant the elevation of the vice president, Tyler, a native of Virginia. Texas, which had seceded from Mexico in 1836, was annexed to the United States near the end of his administration (1841–1845), and formally admitted on July 4, 1845. The Texas issue defeated Clay's hopes of the presidency in 1844 and weakened the Whig Party.

J. K. Polk (1845–1849), a native of North Carolina, the first dark horse ever nominated for the presidency, aggressively pressed for settlement of the Oregon boundary dispute under the slogan "Fifty-Four Forty or Fight" and secured recognition of a boundary in 1846. This aggressiveness was designed to increase the number of states in the north, to parallel the increase in the south with the addition of Texas and the acquisition of New Mexico and California. Americans in California took a hint from Polk and declared an independent state. Polk ordered General Zachary Taylor to occupy the left bank of the Rio Grande; at length the exasperating Mexicans committed an overt act, which was followed by a brief successful war. In 1847, in "the spot" resolutions, Lincoln took an active part in attacking Polk, and to a resolution of Congress thanking General Taylor, secured the addition of a rider that the war had been started by Polk "unnecessarily and unconstitutionally."[19] Polk[20] was accused by the Whigs of forcing a war to extend the institution of slavery. Opposition to the aggressiveness of the south in the interests of new territory became more vocal through the activities of Lincoln and organs such as the *Chicago Tribune*.

Again to capture the electorate, Thurlow Weed, a skilful journalist and politican, played an active role in securing the selection of General Taylor, a native of Virginia, and the hero of Buena Vista (February 1847). He was selected at the Iowa convention within a month of his victory and later triumphantly elected. Vice President Fillmore, a native of New York, became President on his death in 1850 and like most vice presidents not in harmony with the policy of the administration, reversed it. He was sympathetic to the South, and made the first effort of a president to purge his party by opposing the nomination of Whig congressmen who had voted against the Clay compromise.[21] In 1852 the Whigs nominated Winfield Scott, the general who had led the troops to Mexico City, but he was defeated by Pierce (1853–1857). Newspapers exploited such remarks of Scott as "I never read the *New York Herald*" and "the hasty plate of soup."

The long struggle between the North and South was drawing to a close as the North was no longer able to offset southern influence by such tactics as nominating generals for President. These tactics had been to an extent self-defeating since military power was reinforced by recognition of heroes in elections to the presidency. The Whig Party[22] was replaced by the Republican Party supported by the free soil movement. The plantation system led to the acquisition of Indian and Mexican lands. The spoils of Mexico were poisoning the political system—each addition of territory accentuated the rivalry between North and South. The gold rush in California precipitated a more intense struggle for control over the first transcontinental railway. Jefferson Davis, Secretary of War under Pierce, a native of New Hampshire and a minor national hero at Buena Vista, insisted on a Pacific railway along the Mexican border linking California to the Gulf states and opening the trade of Asia to the plantation society. In the north, on the other hand, Stephen Douglas of Illinois demanded a route through Nebraska.

Mastery of the South was evident in the nomination and election of weak northern presidents—Pierce and Buchanan (1857–1861), the latter with the advantage of having refused to wear court dress in England,[23] and the distinction of being the only president from Pennsylvania. Compromises between North and South included the reciprocity treaty with the British colonies in 1854 designed to extend the influence of the North as a balance to expansion in the South. Finally the Supreme Court reflected the influence of the South when it appeared as an agent for southern expansion in the Dred Scott decision. The nomination of Lincoln from the Middle West by the Republican Party and his election brought southern expansionism to an end. Robert E. Lee, a contemporary of Jefferson Davis at West Point, became in 1865 General-in-Chief of the Confederate armies. Withdrawal of able generals to the southern armies compelled the North to build up the effectiveness of a widely separated staff, with activities co-ordinated through the telegraph; the attempt was eventually successful under Grant. Inefficient military leadership in the North meant a longer period of war, greater loss of life, and greater bitterness toward the South. After the savagery which characterized Sherman's march through Georgia to the sea, reflected in his remark "War is hell," the prospects of reconciliation were slight. A revival in the Civil War of the savagery of ideological warfare established precedents for the twentieth century.

At the end of the Civil War a national army had emerged to serve a national state. The President and executive were supreme above the states. Washington became the significant capital and state governments became less important. The South was invited to join a vastly different union than that she had left, but in turn the war had created a solid and a different South from the one which had left the union. Ideological warfare had been carried to great length. The North imposed a peace more bitter than war. The Republican Party, as a result of the costs of civil war and victory, became a sacred cause to New England, the farmers of the Middle West, veterans concerned with pensions, and negroes. Andrew Johnson (1865–1869) was finally disregarded as President. In spite of the Constitution, the President was deprived of control of the Army and governments in the South which had been elected in 1865 were replaced in 1867 by military rule with the whole area divided into five military districts each under a major general. Grant, trained as a general, became the head of an executive which had been built up by a skillful politician but which had deteriorated under Johnson who followed the precedent of vice presidents in reversing policy. Like Jefferson Davis, Grant carried the dominating qualities of a soldier into the administration of civil affairs (1869–1877). He was thwarted in his ambition to annex San Domingo in the south by Sumner, chairman of the Foreign Relations Committee of the Senate, who long served as a focus of northern bitterness, following the savage physical attack on him by Brooks of South Carolina on the floor of the Senate,[24] and who insisted on the acquisition[25] of Canada to the north.

With the aggressive support of Union veterans of the Grand Army of the Republic, Hayes, a brigadier general under Sheridan, was elected to the presidency by a narrow margin in 1876 (1877–1881). In his fight with the Senate, the telegraph became an effective instrument in the mobilization of public opinion. He acquired control of the appointive power and "the long domination of the executive by the Congress was at an end" (H. J. Eckenrode). Grant had been unable to restore the South to white rule because of the Army and the bitterness following the war but under Hayes, as a result of the cohesiveness of white southerners in the Democratic Party, the retreat of the North from the South was begun. It was finally ended in 1894 and the negro was left a third-class citizen, legally free, but deprived of his vote. On the other hand Hayes began the unfortunate precedent of using his power over federal troops to break strikes in West Virginia, Pennsylvania, and Maryland.

Hayes was followed by James A. Garfield, a brigadier general at Shiloh, who to become President defeated General Winfield Scott Hancock, a Union commander at Gettysburg, "a good man weighing 280 pounds" (W. O. Bartlett, in the *Sun*). Garfield, supported by Whitelaw Reid of the New York *Tribune*, had defeated Conkling and the New York *Herald* in the attempt in 1880 to nominate Grant for a third term.[26] The appointive powers conceded to Hayes led to a concern with the introduction of civil service reform but since domination of the Senate necessitated a rigid control over patronage, a strict merit system was impossible. Factors responsible for the murder of Lincoln, vicious personal bitterness, the war, disappearance of an interest in great causes, and the growth of the spoils system culminated in the assassination of Garfield,[27] the defeat of Blaine and the election of Cleveland, and the return of the Democratic Party. (Arthur, Vice President under Garfield, became President in 1881, but contrary to the usual practice did not change his policy.)

On its return to power in 1885 the Democratic Party and its President, though relatively free from the hatreds exploited by the Republican Party, was inexperienced and undisciplined. A forceful leader, Cleveland (1885–1889, 1893–1897) strengthened further the position of the executive in opposition to the Senate. He was defeated by his tariff message of December, 1887, and by Benjamin Harrison (1889–1893),[28] a grandson of President William Henry Harrison elected in 1840, a great grandson of a signer of the Declaration of Independence, and the last of the aristocrats in American politics, and a brevet major at the end of the Civil War. The unpopularity of the McKinley Tariff and the depression contributed to the re-election of Cleveland as president in 1892. Inexperience and lack of discipline in the party, and continuation of the depression were to defeat him. Neglect of monetary reform and an emphasis on the tariff, incidental to the revival of southern influence, led to Bland's warning to Cleveland, in the "parting of the ways speech" in 1893, and a breach between eastern and western Democrats. The weakness of Cleveland in the party was not unrelated to various tactics designed to strengthen his position as President. Although a Democrat he followed the precedent of Hayes in sending federal troops to stop the Pullman strike in Chicago and destroyed the last vestiges of state sovereignty which had maintained the safety of commerce depended on the power of the state.[29] Richard Olney,[30] his Secretary of State, held "any permanent political union between a European and an American state unnatural and inexpedient"—a statement of interest to Canadians. He sent instructions of an inflammatory nature to the American minister in London regarding the dispute between Great Britain and Venezuela, and Cleveland sent a message to Congress which revived feelings of antagonism against Britain. The Navy was rehabilitated and Mahan's writings on naval power developed as an important influence.

The vigorous note to Great Britain was designed to attract Irish votes since the Democratic Party in the North had been built up around the Irish American element in New York State.[31] The words "Rum, Romanism and Rebellion" used by a supporter of Blaine had contributed to the latter's defeat in 1884.[32] In turn the outcome of the election of 1888 had been influenced by a letter which Sackville-West, British Minister in the United States, was tricked into writing to the effect that the interest of Great Britain would be best served by the return of Cleveland.[33] In that election the charge of subservience of the Democratic Party to the Southern Confederacy had been heard for the last time. In 1896 the free silver campaign of the West drove the gold standard Democrats in the East out of politics and weaker elements of the party came to the surface.[34]

As a nominee of the Democratic Party reflecting the demands of the West for monetary reform, Bryan was defeated by W. J. McKinley (1897–1901) who had served as a private, and was a brevet major at the end of the Civil War. The war mania, developed over the Venezuela dispute, persisted and led to demands for war with Spain. This Congress declared in April, 1898. "McKinley had in part given in to public pressure, for fear of disrupting his party and losing the autumn elections."[35] "From the Rio Grande to the Arctic Ocean there should be but one flag and one country!" was the cry of Henry Cabot Lodge. Regarding the Philippines, McKinley decided that "there was nothing left for us to do but take them all, and to educate and uplift and civilize and christianize them," a process involving a long period of hostilities with the Filipinos.[36] The Hawaiian Islands were annexed, partly because they would be needed to defend the Philippines. In the peace treaty Puerto Rico was ceded by Spain.

During the war in Cuba, Theodore Roosevelt, God's gift to newspapermen, who had raised the Rough Riders, and, with the assistance of Richard Harding Davis as war correspondent, secured important space on the front pages of newspapers, became a centre of attention.[37] He was elected Governor of New York State, became Vice President in McKinley's second term, and President (1901–1909) on the latter's assassination. This was attributed to an incendiary press, particularly the writings of Bierce and the Hearst papers, which supported the Democratic Party.[38] Such was the background for a belief in power for the central government; "I achieved results only by appealing over the heads of the Senate and House leaders to the people, who were the masters of both of us."[39] Cleveland gave out messages on Sunday evenings[40] to get more space in the Monday papers and Roosevelt exploited the practice following the development of Sunday papers by making important statements on Sunday and compelling the dull Monday papers to feature them.[41] He prepared speeches well ahead of time in order that they could be distributed to all newspapers before public delivery and the expenses of telegraphing them be avoided.[42] The interest of newspapers in his activities was a result of his sense of news, and of his concern with trust busting, which implied defeat of the International Paper Company as a trust, and lower prices of newsprint. "I took the canal zone and let Congress debate." The Panama had "a most just and proper revolution."[43] In spite of Congress he sent the United States fleet to the Pacific to impress the Japanese. Under pressure from Roosevelt the Canadian claim in the Alaska boundary dispute had been sacrificed.[44] Regarding the appointment of judges to the Supreme Court, Roosevelt wrote: "he [a judge of the Court] is not in my judgment fitted for the position unless he is a party man, a constructive statesman. . . ."[45] His position was summed up in his statement: ". . . I did greatly broaden the use of executive power."[46]

In 1909 W. H. Taft, the nominee of President Roosevelt and the Republican Party, became President (1909–1913). He had been Governor of the Philippines from 1900 to 1904, Secretary of War after 1904 when he successfully reorganized work on the Panama Canal and was described as "an amiable island, completely surrounded by men who know exactly what they want." He attempted to secure the passage of a reciprocity treaty in 1911 but the attitude of President Roosevelt in the Alaska boundary dispute had done much to stimulate hostility leading to its defeat in Canada. The increasing power of the executive, following Hayes and Cleveland, was accompanied by the emergence of the Speaker as an important channel between the executive and Congress. T. B. Reed became the Speaker in 1889, when the Republicans captured both houses and the presidency; a continuous representative from Maine, he was responsible for a marked increase in the importance of the position. The weakness of the Democratic Party, and the position of the Speaker, first in the case of Reed and then in the case of Cannon, in the Republican Party, precipitated a revolt in the latter party in 1910. After that date, the Speaker was excluded from membership in the Rules Committee of the House and lost his power to appoint its Standing Committees. As a result the President had no one person with whom he could deal, and bitterness between factions of the Republican Party led to the emergence of ex-President Roosevelt with a Progressive Party and the election of President Wilson in 1912.

The election of President Wilson (1913–1921) was not only a result of the difficulties of the Republican Party but also of the steady improvement in the discipline and solidarity of the Democratic Party. Champ Clark's blunder in coining a phrase which was used with such telling effect in Canada against the reciprocity treaty in 1911 helped to defeat him as a nominee of the Democratic Party.[47] Woodrow Wilson was a native of Virginia, and his election, first as Governor of New Jersey, and then as President, pointed to a return of southern influence in the Democratic Party. The long period in the wilderness was followed by aggressive legislation in the fields of both tariff and monetary reform. In Wilson's second term, begun with a narrow majority, patronage played an important role in maintaining the discipline of the party. After the outbreak of war, Wilson, according to Lindsay Rogers, became King, Prime Minister, Commander-in-Chief, party leader, economic dictator, and Secretary of State for Foreign Affairs. In the words of Josephus Daniels, "My party has the responsibility of this war." Exclusion of Republicans from the peace delegation meant that Wilson's promises became party politics.

The overwhelming burdens of the war on the executive took their toll in the breakdown of the President's health, in the defeat of the League of Nations by Congress, and in the nomination of Warren Harding from Ohio (1921–1923), "the fine and perfect flower of the cowardice and imbecility of the Senatorial cabal that charged itself with the management of the Republican convention" (*New York Times*).[48] Colonel George Harvey had played an important role in the election of Wilson but the latter feared the possible charge of support by New York interests, especially J. P. Morgan and Co. That his fear was justified is evident in the fact that he was given the nomination by the Democratic Party partly as a result of Bryan's attack on Champ Clark's reliance on New York support. The alienation of Harvey by Wilson was followed by his aggressive interest in the election of Harding and by his appointment as Ambassador to the Court of St. James.[49] Roosevelt had regarded settlement of the Irish question as "most essential to the furtherance of friendship between America and Britain"[50] and Harvey took an active part in establishing the Irish Free State and weakening support of the Irish vote to the Democratic Party. He was instrumental in carrying out the views of the British and Americans in bringing to an end the Anglo-Japanese alliance by a four power treaty.

The death of Harding in office meant the elevation to the presidency of Coolidge from New Hampshire (1923–1929). The religious issue was important in the defeat of Al Smith[51] as it had been in the defeat of Seward[52] by Lincoln at the Republican convention in Chicago in 1860. The defeat of Hoover (1929–1932) was in part a result of the jealousy of correspondents of the preferred position given to one of their number, Mark Sullivan, the difficulties of developing effective relations with the press in various administrative departments, and exploitation of this fact by Charles Michelson in a smear Hoover campaign. Libel laws were avoided by resort to the privileges of the *Congressional Record*.[53]

The disastrous results of the bitter aftermath of the Civil War shown as late as in the uncomfortable position of President Wilson and the attitude of the Republican Party toward the peace treaty, were ultimately evident in the successive readjustments of the terms of peace, in the collapse of 1929, and the election of President F. D. Roosevelt, formerly Governor of New York. He exploited to the full the systematic efforts of Theodore Roosevelt to rid the name of association with the aristocracy.[54] Extensive control over patronage, the advantage of radio in appealing to the people over the head of Congress, and the disciplined support of labour enabled him to dominate the party until his death and enabled the party to dominate Congress to the present. "The radio . . . the supreme test for a presidential candidate" was Roosevelt's "only means of full and free access to the people."[55] He was extremely sensitive to public opinion especially the opinion of religious groups.[56] The picture changed from one of a little-regarded presidential office and a supreme legislative branch under Harding, Coolidge, and Hoover and the strong position of business interests represented by lobbies, to one featuring a strong executive and a vast patronage to executive agencies.[57] In 1938 enormous relief funds were shifted toward preparation of armaments.[58] Even the Supreme Court which, as Chief Justice Hughes remarked, says what the Constitution is, generally sympathetic to the legislative branch of government, after a bitter struggle[59] became more sympathetic to the executive. Finally the transfer of the Bureau of the Budget from the Treasury Department gave the President access to all activities of the government.

The disequilibrium created by a press protected by the Bill of Rights had its effects in the Spanish American War, in the development of trial by newspaper, and in the hysteria after the First World War. Holmes wrote "when twenty years ago a vague tremor went over the earth and the word socialism began to be heard, I thought and I still think that fear was translated into doctrines that had no proper place in the Constitution or the common law." The effects of this hysteria were registered in the influence of the press on legislatures and on the Supreme Court (notable dissents only prove its strength). As a result power shifted increasingly to the executive and involved reliance of the executive on force. In the words of Brooks Adams: "Democracy in America has conspicuously, and decisively failed in the collective administration of common public property."

The power of the President in his control over patronage and party was not only enhanced by the radio but also by military considerations. The importance of the military factor strengthened the possibilities of leadership by a single person with power to intervene in war in spite of public opinion and of Congress. He was compelled to exercise wide discretion to lead or to force Congress to recognize and to accept his power and position. The position of the Democratic Party and the President in the First World War, and in the Second World War, particularly as a result of the radio which widened the gap between the executive and the legislative branches, made it necessary to rely on important intermediaries—House in the case of President Wilson and Hopkins in the case of President F. D. Roosevelt.[60] In Great Britain by way of contrast the Prime Minister had the support of coalition and of Parliament. The solidity of the parliamentary tradition made it possible to defeat and to re-elect Churchill whereas the continued dominance of the Democratic Party, while facilitating the transfer of power from Roosevelt to Truman, meant that changes could only be made in personnel, including members of the Cabinet. Americans were amazed at the necessity of Churchill's maintaining constant touch with the British Cabinet in drawing up the Atlantic Charter in Newfoundland in contrast with the independence of Roosevelt.

In the conduct of foreign affairs, a lack of continuity,[61] incidental to the importance of individuals, and in spite of the encouragement given to careermen in the Rogers Act of 1924,[62] was in strong contrast with the continuity evident in Great Britain and in Russia. This made for less attention to Europe, especially since the importance of interests in Latin America meant greater concern with ministers from these countries, particularly as they were men of ability and industry.[63] Difficulties in conducting negotiations with English representatives were evident at Bretton Woods, Washington and Savannah. English negotiators were constantly faced by Americans with the statement that they could not get that through Congress. The judgment of

American negotiators as to the political tolerance of Congress and of public opinion became a determining consideration.

III

The conflict between Cavalier and Roundhead, between absolute monarchy and absolute parliament, in England was transferred to North America. The southern colonies established at an earlier date reflected the influence of aristocratic organization and the northern colonies the influence of Puritan organization. The demands of the northern colonies for independence with relation to trade were paralleled by demands of the southern colonies for independence in relation to land. In the Revolutionary War the experience of George Washington in the colonial wars with the French became the basis for his appointment as military leader and in turn as President for two terms. He was followed by John Adams, a representative of New England, for one term. From 1801 to 1825 the three Presidents, Jefferson, Madison, and Monroe, each with two terms, were natives of Virginia. John Quincy Adams from New England served for one term and Andrew Jackson, a native of South Carolina, for two terms. He was followed in 1837 by Van Buren of the same party, the first President to be chosen from the middle colonies, who served one term. By this time the middle and northern colonies had built up the Whig Party and succeeded by emphasizing military prestige in securing the election of General W. H. Harrison, followed by Tyler, a native of Virginia. The latter was followed by J. K. Polk, a native of North Carolina, and nominee of the Democrats. The Whigs nominated another military hero, while his laurels were still green, General Taylor, a native of Virginia, and again secured his election. In 1852 and in 1856 the Democrats succeeded by nominating weak northern Presidents, Pierce and Buchanan. Before 1861 all but two of fifteen administrations represented the Democratic Party and of the thirteen nine were served by southern presidents. The Jefferson revolution from 1800 to 1860 was followed by Republican policy from 1860 to 1932.[64]

The dominance of representation from the South and especially Virginia, and of representation from the Army in the period prior to the Civil War, was a reflection of the dynamic power of the plantation system and its demand for more and better land. The weakness of the Spanish, Indians, and Mexicans made it possible for an aggressive government to steadily expand its territory to the west. Expansion of territory to the southwest gave an impetus to parallel expansion to the northwest to be accomplished with an occasional extension of territory at the expense of the British, for example in

Maine and Oregon, and at the expense of the Russians on the north Pacific coast. In the race for land to the west and with its disappearance, the South attempted to expand territory for the slave trade along the northern border of the southern states. The friction eventually led to the outbreak of civil war or the war between the states.

With the end of the Civil War presidents were elected from the North and were again largely representative of the successful northern army. The aggressiveness of the North was checked by growing nationalism in Canada evident in controversies, over the fisheries centring around the Washington Treaty, the Alaska boundary dispute, and the reciprocity treaty of 1911. It took new forms in a continuation of the war against Spain and was effective in the addition of new territory.

Broadening of the powers of the executive such as those boasted about by Theodore Roosevelt and the improvement of communication notably in radio strengthened the position of the President. Control over vast sums following the depression and continued during the war enabled the President to control the party. The seven principles of politics, five loaves and two fishes, were handled more effectively. Patronage and assistance in elections were distributed in accordance with the record of the roll calls in Congress.[65] In the election of presidents directly by majority vote was registered the importance of the middle class urban vote, especially of New York, and the election of senators, following the abolition of election by caucus,[66] two from each state representing predominantly a rural middle class, increased the possibilities of friction.[67] The House of Representatives also reflected the influence of the urban vote but its size left it exposed to vicious partisan and predatory interests and to manipulation under stupid rules such as prevailed under Cannon and after 1925 under the Longworth Snell Tilson triumvirate.[68] It has been described as the greatest organized inferiority complex in the world.

With the tendency toward increased power in the executive and the increasing importance of urban centres the policy of parties is less dependent on a single figure in the presidency. Family names will probably persist as a factor in the selection of presidents—to mention Harrison, Roosevelt, and Taft—and the dangers of assassination[69] will be checked by strengthening of the secret service. Formerly vice presidents were selected as representatives of a defeated minority within the party and were consequently in a weak position when they rose to the presidency.[70] From 1800 to 1900 only one vice president, Van Buren, was elected in his own right to the presidency.[71] More recently the Vice President has become a regional representative intended to support the President as a representative of a densely populated state. Garner from Texas supplemented Roosevelt as did Wallace from Iowa and Truman from Missouri. Since 1900 three Vice Presidents have been elected in their own right: Roosevelt, Coolidge, and Truman.

The importance of New York State and of the possibilities of rapid advance in political life by attacks on corruption explained the prominence of Tilden who attacked the Tweed ring and as Democratic candidate opposed Hayes; of Cleveland who made his reputation in Buffalo; of T. R. Roosevelt, who was New York police commissioner; of Charles Hughes, Republican candidate in opposition to Wilson, who came into prominence in the insurance investigation; and of Dewey with his prosecutions. The intensity of the struggle in New York[72] was evident in the efforts of Hearst to become mayor and governor and eventually president. Coolidge emerged as a national figure in the Boston police strike. Perhaps the comparatively healthy state of New York in spite of the scale of its problems has been partly a result of its possibilities in the making of reputations by attacks on corruption.

The President cognizant of his power must be constantly alert to the implication of policy for voting strength. In foreign policy the results have been evident in several directions. Timing has been carefully worked out in relation to voting or rather voting has been carefully planned in relation to time. A rigid time arrangement compels an emphasis on maneuverability or the settlement of issues when the effects will be most evident in relation to votes. Mr. Truman immediately before the election in 1948 decided to recognize Palestine and to strengthen the position of the Democratic Party in New York State of which Mr. Dewey was Governor. A period of tension and war enormously increases the executive power. The opposition is prevented on the large vague grounds of security and military secrecy from discussing effectively the most crucial element of policy. During the war Republicans were appointed to the Cabinet and bi-partisan responsibility in foreign affairs was assumed. The argument about swapping horses in midstream has proved difficult to answer. It might be answered by nominating a general, let us say Eisenhower, but West Point has never produced good politicians, and he may be content with actually having more power than the President. F. D. Roosevelt, with a personal interest in the Navy, left Army experts with much greater freedom of decision.[73] Such freedom, however, tends to throw the President into the hands of the armed forces. The two-thirds rule regarding treaties in the Senate has been effective in checking the foreign policy of presidents and has been exploited by German, Russian, and Clan-na-Gail delegations,[74] but it has been of little avail with the development of the United Nations and the power of armed force. Indeed the Senate has shown considerable readiness at the demands of the party to co-operate with armed forces.

In the twentieth century the enormous development of industry accentuated by war has greatly enhanced the problems of the executive. Use of the blockade and the threat of blockade has increased dependence on domestic industries. "An all-round increase in armed forces" has been necessary "to mitigate unemployment." We must have "war to solve unemployment in order to ensure against internal anarchy, instead of war solely to protect employment (ordered life) against external aggression." "The dependence on war has become even more vital to our economic system than the dependence of war on industry." "Should an enemy not exist he will have to be created."[75] "A war cannot be carried on without atrocity stories for the home market."[76]

IV

These remarks have been made by one who does not pretend to understand the United States and who cannot appraise the significance of the party struggle as part of the domestic scene. But we are required in the interests of peace to make every effort to understand the effects not only of the actions of the United States but also of our own actions. We have never had the courage of Yugoslavia in relation to Russia and we have never produced a Tito. We have responded to the demands of the United States sometimes with enthusiasm and sometimes under protest. Members of the British Commonwealth struck back against the Hawley-Smoot tariff in the Ottawa Agreements. But we have been a part of the North American continent. The enormous increase in the production of wheat on this continent in the last century was directly related to the Russian revolution, the rise of agrarianism in Germany, of higher tariffs in France and of marked adjustments in England. Germany imposed a tariff on sugar to secure independence in supplies of sugar, drove down the prices of cane sugar, contributed to the outbreak of revolt in the Spanish American colonies, and enabled the United States to take full advantage of the break-up of the Spanish empire.[77] The immigration quota of American legislation in 1924 accentuated the population problems of Italy and contributed to fascism. The silver purchase agreement of 1934 and the consequent destruction of the Chinese monetary system were related to the revolution in China. The protectionist policy of North America and the difficulties of penetrating the American market compel the United States to export dollars and at the same time make it difficult for other countries to acquire dollars. As a result there is resort to enormous expenditure on armament. In the words of the late Carl Becker, what we didn't know hurt us a lot.

A written constitution with its divisive nature established by the Declaration of Independence and the Constitution, centralization under Washington and Adams, decentralization from Jefferson to Lincoln, and centralization after Lincoln, first under the Republican Party and later the Democratic Party, so that at one time there has been a weakening of the power of the executive and at another a strengthening of that power depending largely on the dominant medium of communication, stand in sharp contrast with the unwritten constitution of Great Britain and the undivided power of the Prime Minister responsible to Parliament. In the United States parties are "devoted to the search for compromise between sectional, class, and business groups" and are "frankly uninterested in logical programs or 'eternal' principles."[78] The practice of representation from party rather than regions characteristic of Great Britain finds no expression in the United States.[79] "The most profound of American political thinkers saw in the perpetual search for compromise between selfish interests the basic principle of free government." In the words of Calhoun, "the negative power . . . makes the constitution—and the positive . . . makes the government. The one is the power of acting—and the other the power of preventing or arresting action. The two, combined, make constitutional government."[80] The emphasis on negation, the constant fear of Leviathan, of the encroaching state, has been offset by the promotion of strong government by war and industrial revolution.[81] Under the American Constitution reliance on force has become increasingly necessary whereas under the British, following the brief period in which Parliament was dominated by Cromwell and the army and the period in which the Duke of Wellington was Prime Minister, force has been increasingly subjected to the authority of Parliament. A general as Prime Minister of England would be unthinkable, though the influence of the army and navy are not to be disregarded, whereas in the United States a general as President has been regarded almost as a rule. Ostrogorski has quoted the remark that God looks after little children, drunken men, and the United States. I hope it will not be thought blasphemous if I express the wish that He take an occasional glance in the direction of the rest of us.

Notes

1. Read at a meeting of the Salmagundi Club on December 6, 1951.
2. E. D. Martin, *The Behavior of Crowds: A Psychological Study* (New York, 1920), p. 223.
3. Ibid., p. 248.
4. Ibid., p. 249. "The most certain test by which we judge whether a country is really free, is the amount of security enjoyed by minorities." "By liberty I mean the assurance that every man shall be protected in doing what he believes his duty, against the influence of authority and majorities, custom and opinion. . . . It is bad

to be oppressed by a minority, but it is worse to be oppressed by a majority" (Lord Acton). See Sir John Pollock, Bt., *Time's Chariot* (London, 1950), pp. 166–67.

5. J. F. C. Fuller, *Armament and History* (London, 1946), p. 101.

6. Ibid., p. 109.

7. Herbert Agar, *The United States: The Presidents, the Parties and the Constitution* (London, 1950), p. 28. "For it is a fact, that more than one third of their general officers have been inn-keepers, and have been chiefly indebted to that circumstance for such rank. Because by that public, but inferior station, their principles and persons became more generally known" (Smyth; cited by Kittredge, *The Old Farmer and His Almanack* [Cambridge, Mass., 1920], p. 264).

8. Agar, *The United States*, p. 37.

9. Ibid., p. 45.

10. Ibid., p. 88.

11. Washington, of course, in his Farewell Address had said, "It is our true policy to steer clear of permanent alliances with any portion of the foreign world, so far, I mean, as we are now at liberty to do it."

12. Cited in Agar, *The United States*, p. 174.

13. Ibid., p. 200.

14. James Cheetham, an exile from England after the Manchester riots in 1798, attempted in the *American Citizen*, a daily sponsored by Clinton, to break the power of Aaron Burr in New York. William Duane, editor of the powerful Jeffersonian paper, the *Aurora*, because of a bitter grudge against Madison and Gallatin who refused to give him a job contributed to the defeat of the Navigation Act of Gallatin and hastened the outbreak of war.

15. This had been the *Independent Gazetteer* of Philadelphia under Joseph Gales, a son of the editor and proprietor of the *Sheffield Register*, who had left England following a charge of sedition in 1795. It was purchased by S. H. Smith in 1800 and moved to Washington.

16. A. K. McClure, *Recollections of Half a Century* (Salem, 1902), pp. 37–39.

17. J. E. Pollard, *The Presidents and the Press* (New York, 1937), p. 147.

18. The anti-Masonic party put Seward in the New York State Senate in 1830, made Joseph Ritner Governor in Pennsylvania in 1835, and supported an alliance of J. Q. Adams, William Wirt, Francis Granger, and Thurlow Weed. It carried Vermont for Wirt and Ellmaker, candidates for President (C. T. Congdon, *Reminiscences of a Journalist* [Boston, 1880], p. 29).

19. See R. S. Harper, *Lincoln and the Press* (New York, 1951), p. 9.

20. T. W. Barnes, *Memoir of Thurlow Weed* (Boston, 1884), p. 172.

21. H. L. Stoddard, *Horace Greeley, Printer, Editor, Crusader* (New York, 1946), p. 149.

22. The Whigs failed to capture the popular vote. Daniel Webster was alleged to have said that they should "come down into the forum and take the people by the hand," words which were printed innumerable times in the largest type in Democratic newspapers. Governor J. A. Clifford, on the other hand, imprudently called

the Democrats "poor in character and meager in numbers" (Congdon, *Reminiscences of a Journalist*, p. 61).

23. Pollard, *The Presidents and the Press*, p. 293.
24. See Congdon, *Reminiscences of a Journalist*, p. 253.
25. *The Education of Henry Adams: An Autobiography* (Boston, 1918), p. 275.
26. Pollard, *The Presidents and the Press*, pp. 480–86.
27. Agar, *The United States*, p. 533.
28. J. S. Clarkson, assistant Postmaster-General, a former teacher and journalist, is said to have distributed 38,000 post offices and to have secured the election of Harrison in opposition to Blaine (Herbert Quick, *One Man's Life* [Indianapolis, 1925], p. 220).

> In numerous instances the post-offices were made headquarters for local party committees and organizations and the centers of partizan scheming. Party literature favorable to the post-masters' party, that never passed regularly through the mails, was distributed through the post-offices as an item of party service, and matter of a political character, passing through the mails in the usual course and addressed to patrons belonging to the opposite party, was withheld; disgusting and irritating placards were prominently displayed in many post-offices, and the attention of the Democratic enquirers for mail matter was tauntingly directed to them by the post-masters. (Cleveland, cited by Agar, *The United States*, p. 550)

29. McClure, *Recollections of Half a Century*, p. 131.
30. He threatened the *World* with application of a statute of January 30, 1799, in complaint of its influence on the conduct of Great Britain in relation to the Venezuelan controversy (J. L. Heaton, *The Story of a Page* [New York, 1913], pp. 112, 122).
31. W. J. Abbott, *Watching the World Go By* (Boston, 1933), p. 74. J. Y. McKane, a Coney Island boss, failing to secure benefits from Cleveland, became very active in opposition to him (James L. Ford, *Forty-Odd Years in the Literary Shop* [New York, 1921], pp. 345–46).
32. As Thomas Nast had done effective work as a cartoonist in the election of Grant, Bernard Gillam particularly with "The Tattooed Man" in *Puck* was effective in his support of Cleveland. Ford, *Forty-Odd Years*, p. 299. Conkling's refusal to support Blaine in the words "I am not in the criminal practice" gave weight to the attack.
33. Abbott, *Watching the World Go By*, p. 103. Cleveland asked for his recall. This probably served as a counter move to a release of a story in England in 1887 of the possible purchase of the Maritimes by the United States which was cabled as a scoop for American papers.
34. J. D. Whelpley, *American Public Opinion* (London, 1914), p. 18.
35. Agar, *The United States*, p. 624.
36. Ibid., p. 625.
37. Commenting on Roosevelt's Rough Riders, Mr. Dooley wrote: " 'Tis 'Th' Biography iv a Hero be Wan who Knows.' . . . If I was him I'd call th' book 'Alone in Cubia'" (Elmer Ellis, *Mr. Dooley's America* [New York, 1941], p. 145).
38. Abbott, *Watching the World Go By*, p. 139.
39. Agar, *The United States*, p. 639.
40. Pollard, *The Presidents and the Press*, p. 517.

41. Abbott, *The United States*, p. 244.
42. Oscar King Davis, *Released for Publication: Some Inside Political History of Theodore Roosevelt and His Times, 1898–1918* (Boston, 1925), p. 102.
43. Agar, *The United States*, p. 650.
44. Ibid., p. 626.
45. Ibid., p. 644.
46. Ibid., p. 638.
47. C. B. Davis, "*The Great American Novel——*" (New York, 1938), p. 146. Josephus Daniels claimed that he would have won with the radio as Hoover did later. See James Kerney, *The Political Education of Woodrow Wilson* (New York, 1926).
48. Cited in Agar, *The United States*, p. 675.
49. See W. F. Johnson, *George Harvey* (Boston, 1929), pp. 286 ff.
50. A conversation with Sir Joseph Ward, Prime Minister of New Zealand, in 1909 (Hon. Sir James Kirwan, *My Life's Adventure* [London, 1936], p. 226).
51. For a striking account of the implications of the Coolidge statement "I do not choose to run" for the final disposition of the Sacco Vanzetti case see *We Saw It Happen*, ed. H. W. Baldwin and Shepard Stone (New York, 1938).
52. Seward had been elected Governor in New York in 1838 with the support of the Roman Catholic Archbishop Hughes and had urged a division of the school fund between Catholics and Protestants with the result that he antagonized the strong American native party in Pennsylvania. McClure, *Recollections of Half a Century*, p. 216.
53. Pollard, *The Presidents and the Press*, pp. 743–45.
54. E. C. Bentley, *Those Days* (London, 1940), p. 198.
55. R. E. Sherwood, *Roosevelt and Hopkins* (New York, 1950), pp. 184, 186–87. Every word in his speeches was judged not by appearance in print but by effectiveness over the radio and careful attention was given to accurate timing in relation to the number of words and the rate of delivery (pp. 217, 297). It is significant that before the radio no pre-eminent orator ever succeeded in reaching the presidency (A. K. McClure, *Our Presidents and How We Make Them* [New York, 1900], p. 88). It might also be noted that Blaine and Tilden were the only men who managed their own campaigns for the presidency and that both were defeated (ibid., p. 312).
56. Sherwood, *Roosevelt and Hopkins*, p. 384.
57. R. G. Tugwell, "The New Deal: The Decline of Government," *Western Political Quarterly*, June 1951, pp. 295–312. For a study of the conflict between presidential and congressional authority over the administration see C. S. Hyneman, *Bureaucracy in a Democracy* (New York, 1950).
58. Sherwood, *Roosevelt and Hopkins*, p. 101.
59. J. Alsop and T. Catledge, *The 168 Days* (New York, 1938).
60. Sherwood, *Roosevelt and Hopkins*, pp. 931–33. Ickes complained in 1940 that Hopkins had "never even attended a county meeting and wouldn't know how to get into one. Now here he is taking over a national convention. It's disgraceful" (J. A. Farley, *Jim Farley's Story: The Roosevelt Years* [New York, 1948], p. 297).

61. The diplomatic corps was an adjunct of the spoils system and the football of politicians. See Whelpley, *American Public Opinion*, pp. 113, 121.

62. See Drew Pearson and R. S. Allen, *Washington Merry-Go-Round* (New York, 1931), p. 140.

63. Ibid., pp. 30, 46.

64. McClure, *Our Presidents and How We Make Them*, p. 21.

65. George Michael, *Handout* (New York, 1935), p. 73.

66. For a criticism of the direct primary see C. J. Stackpole, *Behind the Scenes with a Newspaperman: Fifty Years in the Life of an Editor* (Philadelphia, 1927).

67. A. N. Holcombe, *The Middle Classes in American Politics* (Cambridge, Mass., 1940), p. 104.

68. Pearson and Allen, *Washington Merry-Go-Round*, pp. 217–19.

69. The influence of anarchism and the Colt revolver on the disappearance of apparent dictatorships in business and in governments has never been given careful study. See Emma Goldman, *Living My Life* (New York, 1934).

70. H. L. Stoddard, *As I Knew Them: Presidents and Politicians from Grant to Coolidge* (New York, 1927), p. 123.

71. McClure, *Our Presidents*, p. 25.

72. The Democratic Party in New York State became a political workshop of the United States and leaders throughout the United States after 1925 were urged to organize along the lines of New York, especially in giving women an equal voice on committees. The feminine vote became an important factor in 1932 and in 1936. See James A. Farley, *Behind the Ballots: The Personal History of a Politician* (New York, 1938), pp. 55, 160.

73. Churchill exercised much greater control over the army. See Sherwood, *Roosevelt and Hopkins*, p. 246.

74. Count Cassini, a Russian minister, and Von Holheben, a German minister, appealed successfully through the press to the Senate against presidential policy (*The Education of Henry Adams*, p. 375).

75. Fuller, *Armament and History*, pp. 164–65.

76. Bentley, *Those Days*, p. 184.

77. See Brooks Adams, *America's Economic Supremacy* (New York, 1900), pp. 36–41.

78. Agar, *The United States*, p. vii.

79. The fathers were particularly concerned to avoid the borough system of England. "State law and custom have practically established that a representative must be a resident of the district from which he is elected." See D. A. S. Alexander, *History and Procedure of the House of Representatives* (Boston, 1916), p. 5. As a result the mobility of the ablest individuals has been checked, whereas in England parties have been much more effective in attracting and securing the election of the ablest individuals irrespective of residence.

80. Agar, *The United States*, p. vii.

81. Ibid., p. xiii.

CHAPTER THREE

Roman Law and the British Empire

I

It seemed fitting in a programme of lectures to celebrate the one hundred and fiftieth anniversary of New Brunswick that I should be concerned with a country which has played an important role in the life of the institution, namely the United States. This province was created as a result of strategic plans of defence on the part of the second British empire against the colonies which had rebelled. Nova Scotia was divided into three separate areas, Cape Breton, New Brunswick, and Nova Scotia in order to provide separate nuclei around which defensive measures might be mobilized. Loyalists migrated to New Brunswick and kept alive the memories of hostility to their native land. Christopher Sauer, a prominent figure in the history of printing in Pennsylvania, started the first newspaper in New Brunswick. This university began, as your calendar states, through the interest of Loyalists in the education of their children and, in the words of the memorial in 1785, the "necessity and expediency of an early attention to the establishment in this infant province of an academy of liberal arts and sciences."

James Bryce attempted to throw light on the problems of the British empire by emphasizing parallels with the Roman empire and in particular by suggesting the contributions of Roman law and of common law to the development of the respective empires.[1] At the very period in which Bryce was revising his essays for separate publication in 1914 the British empire was undergoing crucial change. Since that date the development of the Commonwealth in the

Statute of Westminster and the changes in status of Ireland, India, and Newfoundland point to a need for reconsideration.

The name of Bryce will always be associated with the results of the first major change in the empire in his *The American Commonwealth*. The American Revolution was a result of limitations of common law, "that ancient collection of unwritten maxims and customs" (Blackstone) which have been discussed by a large number of English, American, and other scholars. Professor C. H. McIlwain has described the problem of common law[2] in the seventeenth century when Parliament reflected the influence of force in the substitution of the Cromwellian régime for that of the Stuarts. The absolute power of the Tudors was replaced by the absolute power of Parliament, and both were regarded as encroachments on common law. Sir Edward Coke defended the position of common law in the Bonham case in 1610. "When an act of Parliament is against common right and reason, or repugnant, or impossible to be performed, the common law will control it, and adjudge such act to be void." But such limitations were not recognized by Parliament under Cromwell or in the establishment of legal supremacy in the Revolution of 1689. The English colonies in North America had been established in the period before Parliament had assumed this position and were unable to accept its implications. James Otis restated the position of Coke, and the Assembly of Massachusetts on March 2, 1773, refused to recognize the supremacy of Parliament. "We conceive that upon the feudal principles all power is in the king; they afford us no idea of parliament." Great Britain had seen the evolution of the supremacy of Parliament at the expense of common law, and the colonies, determined to protect the position of common law,[3] introduced a constitution designed to check the power of legislative machinery.

It will not be necessary to rehearse the steps taken by Great Britain, and the colonies remaining within the empire, to develop a constitution which would evade the disaster of the first empire. The Maritime provinces succeeded in building a second empire from the wreckage of the first in which responsible government was achieved. The common law came into its own with a recognition in Great Britain of the limitations of Parliament and recognition in the colonies that the elaborate machinery of the United States to protect the common law was unnecessary. In Great Britain the effects of a common law parliament were evident in the Reform Bills and in the extension of the franchise in the nineteenth century.[4] Elements in the constitution opposed to its effective operation were steadily weakened as the House of Commons increased in power at the expense of the House of Lords. Long and bitter struggles characterized this change and still characterize it but the legislation of 1911 definitely brought the power of the Lords to an end. "The House of Commons after putting under its feet the Crown and the House of Lords, has in its turn been put under the feet of the caucus."5

The changes within Great Britain had profound implications for the empire. Indeed the legislation of 1911 was directly linked to the problem of Ireland and the possibility of establishing Home Rule. In defeat the Conservative Party opposed the Liberal Party supported by Irish and Labour members, first in the House of Lords and finally in Ulster. The unsavoury story of how the army joined hands with Ulster leaders and leaders of the Conservative Party, described by Prof. M. J. Bonn6 as the beginnings of fascism in Europe, need not be retold. For the first time Parliament was openly and to some extent successfully defied by force. During the First World War Irish opposition became more determined and led to the Easter rising of 1916 and eventually to the treaty and the Irish republic. A common law parliament had become impossible in the face of obstructionist tactics which developed from the Irish question.

It has been suggested7 that British imperialism succeeded in areas in which native populations were eliminated (as in America and Australia) or in areas in which a bureaucracy could be established (as in India), and failed in areas in which a strong cultural influence dominated garrisons of settlers as in Ireland; but the suggestion overlooks the role of common law. Men trained in common law such as Gandhi were quick to see its possibilities in the protection of the rights of individuals. After his training in London, Gandhi carried on an effective campaign in South Africa on behalf of Indian immigrants, and with the techniques developed in South Africa contributed powerfully to the establishment of India and Pakistan. Common law implied concern with local customs and facilitated the development of the British Commonwealth by peaceful means or by minor rebellions.

With the increasing importance of legislation, particularly after the Reform Act of 1832, lawyers continued to play an important role in Parliament in the making and in the interpretation of statutes. Common law countries favour the election of lawyers as legislators to the exclusion, for example, of journalists, in contrast with Roman law countries which seem to favour the latter. In common law countries the state became a part of customs and traditions and the revolutionary tradition was weakened.8 Marx's withering of the state had reference to Roman law and not common law countries. Common law traditions which made politics a part of law and emphasized the relation of the state to law implied an absorption of energies in politics and a neglect of the cultural development which has characterized Roman law countries. The danger of imposing common law traditions on Roman law countries has been evident in the difficulties of the parliamentary system in those countries.

The implications of the dominance of lawyers are suggested in remarks by Sir Henry Taylor:

> Of law-bred statesmen (if they have had practice at the bar) the peculiar merit is a more strenuous application of their minds to business than is often to be found in others. But they labour under no light counterpoise of peculiar demerit. It is a truth, though it may seem at first sight like a paradox, that in the affairs of life the reason may pervert the judgment. The straightforward view of things may be lost by considering them too closely and too curiously. When a naturally acute faculty of reasoning has had that high cultivation which the study and practice of the law affords, the wisdom of political as well as of common life will be to know how to lay it aside, and on proper occasions to arrive at conclusions by a grasp; substituting for a chain of arguments that almost unconscious process by which persons of strong natural understanding get right upon questions of common life, however in the art of reasoning unexercised.
>
> The fault of a law-bred mind lies commonly in seeing too much of the question, not seeing its parts in their due proportions, and not knowing how much of material to throw overboard in order to bring a subject within the compass of human judgment. In large matters largely entertained, the symmetry and perspective in which they should be presented to the judgment requires that some considerations should be as if unseen by reason of their smallness and that some distant bearings should dwindle into nothing. A lawyer will frequently be found busy in much pinching of a case and no embracing of it—in routing and grunting and tearing up the soil to get at a grain of the subject;— in short, he will often aim at a degree of completeness and exactness which is excellent in itself, but altogether disproportionate to the dimensions of political affairs or at least to those of certain classes of them.[9]

As has been said of many lawyers all facts are to them free and equal.

II

An elementary discussion of the conditions under which lawyers work in practising at the bar may suggest more clearly the important role of the legal profession. The lay-out of the court room emphasizes the power and authority which surround proceedings involving life and death. The bench sets off sharply the position of the judge, and below him come the witness stand, the bar for the opposing counsel, an inner bar for His Majesty's counsel learned in the law, and beyond seats for the public. The tradition of awe inspired by these arrangements, the insistence on the dignity of the court, and the rigid prohibitions against smoking, chewing gum, or other distractions which may include the reading from a manuscript by counsel inspire a concern with the search for truth and justice.

Encroachment on these traditions has been evident in the demand for photographs for the press and in the interest of criminal lawyers in publicity. The court has possibilities of advertisement for young lawyers. Even members of the Supreme Court appear to relish the appearance of their photographs in the press. But while lawyers display a keen interest in the details of crime such as those appearing in the press they tend to dislike specialization in criminal law and to prefer a mixed practice of civil and criminal law. Concentration on criminal law is apt to be thought of as having a deteriorating effect on character and reputation.

Court procedure involves dependence on the oral tradition in eliciting testimony from witnesses who have been placed under oath to give the truth, the whole truth, and nothing but the truth. Facts are determined by examination and cross-examination and re-examination of counsel. Opposition between counsel is designed to check and to produce evidence from which the judge or the jury must decide the case. When evidence has been elicited and established, argument to establish the law suited to the facts follows. Respect will be shown in language and demeanour to the bench, cases of dispute with the bench are prefaced by the words "with great respect." The maxim handed on to young members of the bar, "Never talk down to the bench," reflects the egoism of the bar and the necessity of emphasizing the place of the bench.

The significance of the oral tradition is evident in the possibility of checking extravagant statements made by counsel or by witnesses. With a background of development prior to the spread of reading and writing, the tradition of the importance of oral rather than written evidence has persisted in the procedure of the court and in the jury system. The common law has consequently been responsive to the opinion of all classes of society including the illiterate. This contact has possibly been more effective than that of the church and religion since it is without the elaborate ceremonial and the written scriptures of the latter, though it musters support from religion in requiring testimony sworn on the Bible and may exact severe penalties for perjury. English courts will insist on the appearance of living authorities rather than extracts from text-books written by them on the assumption that an authority may have changed his mind after writing the book. In North America the difficulty of transporting a living authority over long distances has favoured a whittling down of the English rule and increasing reliance on the text.

The advantage of the oral tradition which allows for constant change even during the course of the trial becomes evident in the exposure of weaknesses in evidence and in argument. The character of witnesses is brought out in detail and the role of intent more easily established. In the preparation of cases counsel must study intensively the character of his own and other witnesses and estimate strong and weak points in order to work out satisfactory tactics in presentation. The common law gives great emphasis to character and to the study of character from an objective point of view. Its success is linked to individualism and necessitates a concern with the influence of the state on character and of character on the state. There is danger of forgetting the words of the Lord Chancellor: "Necessitous men are not, truly speaking, free men."[10] "It is precisely because the force of circumstances always tends to destroy equality that the force of legislation should always tend to maintain it" (Rousseau).

In stressing the importance of the oral tradition it is necessary to keep in mind the roles of the written and the printed tradition. In England courts are more jealous of their position and check discussion by newspapers when cases are *sub judice*. The dangers of extravagant publicity become acute for members of the jury may come under the influence of public opinion reflected in the press.[11] A more subtle problem arises with the spread of mechanization in the preparation of reports of court proceedings. Since questions and answers are phrased in relation to a sworn record which may become the basis of consideration and decision by the bench they will tend to blur the sharp impressions characteristic of an oral tradition. A concern with the record implies an interest in a type of question suited to reading and a neglect of the transient impression of the spoken word. The oral tradition is carefully warped in relation to the demands of a written or stenographic record. The tendency to concentrate on the record has an advantage in that it enables the bench to study the case in a dispassionate and objective fashion but a disadvantage in that it enables the bench to delay reaching a decision and perhaps encourages continuance on the bench of men who by age or inclination are reluctant to appreciate the importance of promptness in the administration of justice. But there may be warrant for the remark that truth will out even in an affidavit.

The legal profession in itself has an important influence on the administration of justice. Counsel are constantly alert to the artistic character of work done by members of the profession and are continually engaged in the appraisal of the capacities of fellow-members and of those of their ranks appointed to the bench. The essentially feudal character of the legal profession is evident in references to "my lord," "my friend," or "my learned friend." Yet style has become more prosaic and matter of fact and even conveyancing can perhaps no longer be described as "a jungle of antiquated fooleries kept up by the pedantry and interest of those who profited by it."[12]

Clashes between opposing counsel bring out sharply the competition in ability. Each party appearing before the court is obsessed with its own advancement and becomes extremely critical of counsel in cases of defeat. Courtesies between members of the legal profession temper the acerbities of conflict and impose a severe restraint on bitterness. The appearance of conflict in the courts will meet the demands of parties and permit the courtesies of the profession outside the court. The protection of the courts and the interest of counsel in clients help to ensure that questions of fact even of an embarrassing character will be brought out, but they are limited in their operation by the relative capacities of counsel, reflected in the size of fees, and by the ability of clients to pay fees. Large earnings, indeed, assume an enormous importance in the administration of law. Ability is maintained at the bar and restricted on the bench which is apt to be impressed by counsel capable of securing large fees. There appears to be a tendency for large companies to secure protection in legal counsel and for counsel to be able to win large fees in successfully protecting them. Success will depend on the ability of counsel but also on the size of the legal firm. A large firm acquires enormous resources in the specialized knowledge of its members and its ability to attract energetic and able young juniors. The demands for intense industry can only be met by younger men and explain the general impression of relatively short lives in the profession.

The advantage of the large firm has become more evident with the enormous increase in legislation and in the numbers of digests, indexes, and abridgements of reports of cases. The large amount of printed material has been further increased by the growth of black letter law including text-books, commentaries, and the like. Lawyers tend to become lazy with the increase of indexes and digests, to neglect a reading of cases with thoroughness and system and to demand more indexes. The need for an expensive library and the use of abridgements has tended to favour those earning large fees and thus the commercialistic bent of the profession. The spread of printing weakens the oral tradition. The increasing importance of the written, mimeographed, and printed tradition has been accompanied by a decline in the position of the courts and changes in the character of law.[13]

Executorships of wills have largely gone to trust companies and account collections to collection agencies. A marked increase in the mortgage business of insurance and loan companies has led to specialization and the handling of business by larger law firms. So too corporation work has become highly specialized and has come into the hands of large firms. Practice of law in relation to automobile accidents is carried on by lawyers acting for insurance companies. Income tax law has become the concern of legal specialists who are forced to compete with chartered accountants. Labour law is now a special field. The rise of boards of administrative character has meant a demand for specialists other than lawyers. Law has followed the shift from individualism to collectivism. Able young graduates from law schools are apt to become immediately interested in office work rather than court work to the great disadvantage of courts.

Demands on the legal profession have increased with the specialization which characterizes the Western world. Cases are presented before modern courts involving a mastery of highly technical questions in a wide range of subjects. The expert appearing as a witness, whether accountant, economist, engineer, or doctor must be subjected to intelligent examination and cross-examination involving a mastery on the part of counsel of the particular subject under consideration. Clarence Darrow's practice, "I never ask a question unless I know beforehand what the answer will be," is generally followed. The legal profession must maintain a profound belief in its capacity to master any evidence and to adapt all questions to the demands of the court. Counsel are compelled to concentrate intensively on particular problems and to become obsessed with a knowledge of immediate details. The common law with its emphasis on the oral tradition has perhaps a greater interest in the ascertainment of facts than other legal systems. Facts are more important than principles. Litigious procedure, for example, emphasizes circumstantial evidence in contrast to the inquisitorial procedure of code countries. The importance of the jury system and opposition to the use of hearsay evidence through fear of misinterpretation by the jury stands in contrast with other systems, and involves its own handicaps. For example, a purchase from a department store may be proved more easily by an appeal to the sales clerk than by reference to the more certain evidence of the department store's records.

The advantages of the common law system with its emphasis on facts are evident in a society favourable to the scientific tradition and to industrial development in the sense elaborated by Bacon. They are further seen in the emphasis of a common law society on news. Lawyers share the interest of newspapers in questions of the moment. These advantages assume limitations. Considerations involving continuity in time are rather neglected and the long-term factors ignored. A training in law makes for a brittle, brilliant type of work. Lawyers are compelled to master the intricacies of a case and after its completion to forget it and to master the intricacies of the next case. The memory tends to be neglected, general principles to have limited attraction, and general theory to be ignored. Law is apt to become anything "boldly asserted and plausibly maintained." A neglect of the time problem implies a lack of interest in theoretical problems. In contrast, the Roman law tradition in its concern with principles attracts the highest intellectual ability to the academic field and encourages an interest in philosophical theory and theoretical speculation. In turn it becomes possible to develop an interest in problems of continuity of time, though the late Justice Holmes could write: "People want to know under what circumstances, and how far, they will run the risk of coming against what is so much stronger than themselves, and hence it becomes a business to find out when this danger is to be feared. The object of our study then is prediction of the incidence of the public force through the instrumentality of the courts. . . . Far the most important and pretty nearly the whole meaning of every new effort of legal thought is to make these prophecies more precise, and to generalize them into a thoroughly connected system."[14] "Law is the delimitation, morality the evaluation, of interests" (Korkunov).

The attraction of large fees for able counsel weakens the possibility of their being attracted to the bench or to political life but the bench has become more popular as a result of income tax regulations and a prospect of holidays. It has been pointed out that separation of the barrister and the solicitor in England tempers the effect of finance on the legal profession there and that the combination of the two positions in the solicitor in Canada greatly increases the impact of business and finance on the legal profession. During periods of depression with decline in fees counsel will perhaps turn more quickly to political activity. Dislike of living in Ottawa is accompanied by appreciation of the prestige of provincial supreme courts in provincial capitals. Relative absence of restrictions on age of retirement on the provincial bench as compared with the federal courts enhances the attractions of provincial courts and explains to an important extent the relatively high calibre of provincial appointments. Since the salaries of judges in the provinces are uniform, appointments in the smaller provinces with lower living costs and much less business become extremely attractive.

Consequently lawyers assume an intense interest in politics and premiers have become chief justices of the provinces. Politics are apt to be dominated by lawyers and to be slanted in the interest of lawyers. Appointments to the federal Supreme Court and to the provincial courts are of course subject to restrictions in religion, region, and language. The Province of Quebec, partly because of the importance of the civil code[15] as well as common law, partly because of the demands of the French and English populations, has been given three judges on the federal Supreme Court, and in turn the Province of Ontario is represented by the same number, one of whom must be an Irish Catholic. The Maritimes are represented by one member and the Western provinces by two members. The rigidity of conventions in appointments reflects the power of the legal profession to defend its interests. The domination of the Liberal Party in the House of Commons, the Senate, and the judiciary assumes a monopoly of legal knowledge. The effects of these restrictions will be tested more sharply with the abolition of appeals to the Privy Council and they may well prove to have serious consequences for the success of the federal system of government.

Reluctance to accept appointments on the bench because of the attraction of large fees tends to divide the profession into two groups. Counsel less attracted to the courts recognize the importance of political activity. Their training adapts them to the ruthlessness of political life. It requires assiduous study, skill in debate, and constant appearance in public. They enter Parliament and have a direct effect on legislation through statutes and following a political career receive appointments to the bench before whom practising lawyers must appear. Successful practising lawyers are compelled to interpret legislation prepared by, and to practise before, successful political lawyers. Counsel trained in the common law tradition in Parliament and on the bench are concerned with legislation reflecting a common denominator of public opinion and registering the effects of a training with an emphasis on facts. Legal training which assumes a capacity to ascertain and to master factual presentation ensures that Parliament has at its command an array of ability particularly adapted to its varied demands in the enactment of legislation covering a wide variety of subjects, though a dominant party with a strong civil service may greatly handicap the opposition.

The advantages of legal training shown in the capacity for intense concentration (five legal cases at once are regarded as a maximum for lawyers) and the mastery of facts in a short period of time have been evident in the success of lawyers in political life. The effectiveness of legal prime ministers can be illustrated by reference to Lloyd George, not to mention others nearer home. Lloyd George declared, "I should always feel at liberty to override the findings of any body of experts."[16] Though politicians do not receive pensions, Parliament attracts lawyers since their chances of appointment to the bench are greatly improved by political activity. The hazards of political life for the lay politician and the absence of political pensions accentuate competition among lawyers for the bench or for the Senate. It has been said of the United States Supreme Court, "The court is small, the cream (sometimes not very fat cream) of a profession in which the political impulse is strong."[17]

Traditions of procedure in common law countries emphasizing the oral tradition in the court and in Parliament imply a background unsympathetic to the social sciences with their emphasis on the written tradition. Inclusion of courses in the social sciences in the training of the lawyer, and of courses in law in the training of the social scientist, may contribute to a solution of the difficulty and to a reconciliation between law and the social sciences but on the other hand may weaken the distinctive contribution of each. A legal training permits a rapid shift from the intricacies of one case to those of another but this advantage is offset by an inability to penetrate problems to an appreciable depth; training in the social sciences develops a mastery of complex problems but this advantage is offset by an inability to shift quickly from the intricacies of one problem to another. The long and tedious process of working through the complex problems of the social sciences is in sharp contrast with the swift effective argument needed in the law courts. Cross fertilization quickly reaches a point at which its advantages are followed by the disadvantages of cross sterilization. The type of social scientists acceptable to the courts is marked by the ability to ask questions intelligible to lawyers and to answer questions intelligible to lawyers. This type of social scientist rarely enhances his prestige among his fellow social scientists and appears eventually to lose his prestige even among lawyers, who in turn become contemptuous of the complications of the social sciences. Social scientists concerned with fine-spun abstractions tend to neglect a sense of proportion and the practical matters of fact with which common lawyers are obsessed. Social scientists appearing in common law courts are necessarily concerned with immediate problems and are consequently restricted in the development and application of theory. They tend to become advocates and to reflect the points of view of their employers. The longest purse will produce the best economist. The late Justice Holmes may not have been right in saying that "for the rational study of the law, the black letter man may be the man of the present, but the man of the future is the man of statistics and the master of economics"[18] and that "every lawyer ought to seek an understanding of economics," but he was certainly accurate when he said that "the present divorce between the schools of political economy and law seems to me an evidence of how much progress in philosophical study still remains to be made."[19] It is the function of the social sciences and the bureaucracies to offset the effects of the obsession of common law with nominalism. The hierarchy of the law undoubtedly weakened the ecclesiastical and military hierarchies. It has been influential in the development of an effective business hierarchy which has dangers for the hierarchy of law itself. The place of lawyers in business is strengthened by their status in the courts and the place of lawyers in the courts is strengthened by their status in business.

III

Following these remarks on the character and implications of common law I propose to turn to a discussion of the influence of Roman law in the British empire. The British empire emerged in part as a result of a balance between the oral tradition and the written tradition, between common law and Roman law.[20] The element of Roman law, especially as reflected in canon law, which persisted after the Reformation in England, was gradually reduced in importance in the British empire and the results were evident in the Commonwealth. The divine right of the papacy was replaced by the divine right of Parliament after the rebellion. The submergence of the concept of fundamental law eventually precipitated the American Revolution, and the written constitution of the United States was designed to restore it and to protect its position. Emergence of a federal government with a constitution which gave enormous powers to the courts involved protection to fundamental law but, in protest against the divine right of Parliament, assumed the divine right of the United States. Without a written constitution Great Britain was able eventually to master the problem of empire and to digest the element of Roman law or rather to cast it out into regions which left the empire, for example the United States, or into regions which insisted on independence and autonomy within the empire, for instance members of the Commonwealth. "I have been convinced that a democracy is incapable of ruling others" (Cleon).

The element of Roman law which became more powerful in other parts of the British empire was evident in the insistence of small areas on their autonomy and divine rights,[21] in the emergence of a federal system, and in conflicts over the concept ending in the United States in the war between the states. Temporarily the significance of Roman law was lessened but supremacy in the north reflected the importance of the divine right of union which was essential to effective opposition to the divine right of states. With the return of southern influence through the Democratic Party the principle of divine right in the states was protected in an emphasis on the divine right of the United States expressed in such intangibles as a way of life. The pattern of federal government in the United States was followed by members of the Commonwealth, notably Canada and Australia.

The problem of law in the United States incidental to a written constitution has been evident in the position of the press. In 1737 when Andrew Hamilton secured a verdict of not guilty for Peter Zenger from a jury in the colony of New York and thus established freedom of the press, he made the lawyer, especially the country lawyer, a dominant force in political affairs.[22] The relations between the press and the law became more important as a result of the clause of the Constitution of the United States which states "this constitution and the laws of the United States which shall be made in pursuance thereof; and all the treaties made, or which shall be made, under the authority of the United States, shall be the supreme law of the land; and the judges of every state shall be bound thereby, anything in the constitution or laws of any state to the contrary notwithstanding." This clause "gave the courts of the United States a power possessed by the judicial tribunals of no other country. It brought within the scope of the lawyers business cases of a kind that could arise nowhere else—it married law and politics" (Hon. F. W. Lehman).[23]

The crucial position of the United States Supreme Court in relation to politics has been evident in appointments to the bench and in Mr. Dooley's remark that the Supreme Court follows the elections. Following a period in which the constitutional document was dominant or after the death of Chief Justice Marshall, say from 1835 to 1910, dual federalism, the doctrine of the police force, a taboo on delegated legislative power, the derived doctrine of due process of law, and the conception of liberty as freedom of contract, were worked out. Finally in the theory of judicial review the background of accumulated doctrines greatly strengthened the position of the Court in the field of constitutional interpretation.[24] Improved communications brought flexibility and a more rapid response from legislatures and courts to the demands of public opinion. "In the civil war the courts sanctioned everything the popular majority demanded under the pretext of the war power, as in peace they have sanctioned confiscations for certain popular purposes under the name of the Police power."[25] The Dred Scott case shook the confidence of the United States in the impartiality of the Supreme Court, as did the election of Hayes over Tilden[26] in 1876. Theodore Roosevelt expressed his disappointment in the position taken by Justice Holmes on the Northern Securities Case following his appointment to the Supreme Court by the remark that he had not the backbone of a banana.

In the period after the First World War the suppression tactics of Attorney-General Palmer led Clarence Darrow to remark that he had "very neatly managed to overthrow one form of government by force and violence."[27] One of these was to make it unlawful by the Overthrow Act "openly to advocate by word of mouth or writing the reformation or overthrow, by violence or any other unlawful means, of the representative government now secured to the citizens of the United States."[28] The division between Liberals and Conservatives in the Supreme Court after the Red scare persisted to the Roosevelt administration[29] and an even balance which gave one individual in the Court enormous power led to demands for reform. President Roosevelt held that he had worked closely as Governor of New York with the New York Court of Appeals and was anxious to have new justices in the Supreme Court with whom he could confer on plans for social and economic reform.[30]

Though defeated in his attempt to change the Court, Roosevelt succeeded in securing a balance favourable to reforms. For example in a decision in *United States v. Associated Press* (326 U.S. 20), it was stated: "Freedom of the press from government interference under the first amendment does not sanction repression of that freedom by private interests. . . . Surely a command that the government itself shall not impede the free flow of ideas does not afford non-governmental combinations a refuge if they impose restraints upon that constitutionally guaranteed freedom."

The effect of freedom of the press on the position of the citizen in Congress and in the courts has been disastrous. Mr. Justice Rand in the Sidney Hillman lecture on March 29, 1951, stated: "The lowest infamies of the informer have been challenged by the outbursts on the floors of legislative chambers, and the reflex has been a public paroxysm of hysteria. True freedom of speech has become a mockery: a man's social and economic life may be destroyed by the chit-chat of a cocktail party." In the courts "the lawyer is trained to elicit fact rather than truth. It is suggested that the oath should read 'I solemnly swear to give the best answers I can under the rules of evidence to such questions of fact as the judge allows to be put to me.' "[31] "The requirement that a witness must be limited to facts is a nail in the coffin of the truth."[32] "Our method of getting at the truth through the medium of the human faculties is crude and cumbersome beyond belief."[33]

For the lawyer "the most important thing to do is to make the judge want to decide things your way."[34] Darrow, a successful criminal lawyer, deliberately appealed to public opinion as a means of influencing the judge. As to juries he concluded that old men were more charitable than young men, that Irish and Jews were easiest to move to emotional sympathy, and that wealthy men, Presbyterians and Lutherans were to be avoided.[35] He was concerned to educate members of the jury.[36]

IV

The reaction of the United States and members of the Commonwealth in their attempts to protect fundamental law has left them more imperialistic than the mother country. As we have traced the reassertion of common law in Great Britain and the decline of imperialism we must turn to its decline in the other Anglo-Saxon regions and the rise of imperialism. In the English colonies in North America which became the United States, rights were protected in the constitution. Control over land within the boundary of each state remained in the hands of the state but in the interior of the continent beyond the boundary of the coastal states it was in the hands of federal authorities until a new state was set up and accepted in the union. Expansion across North America proceeded to the Pacific coast and new systems of control were developed beyond the borders in Alaska, Hawaii, the Philippines, and other areas. It has been said that the British empire was acquired in a fit of absent-mindedness, but the empire of the United States has grown up during periods of imperialistic fanaticism marked by such slogans as "Manifest Destiny" and "54-40 or Fight," and during periods when imperialism was thrust upon her as in the Louisiana Purchase. In Canada we have seen American imperialism at work in various ways, ranging from the fisheries disputes to protests against construction of the Canadian Pacific Railway and the duress exercised by President Theodore Roosevelt on the arbitrators in the Alaska boundary dispute. Significantly, other countries are beginning to see the character of American imperialism. American publications protest against appointments of certain cabinet ministers in Great Britain. An American public body passed a resolution demanding the settlement of the Irish question. Shades of George III!

In the United States the shift from an obsession with North American expansion to foreign expansion becomes apparent towards the end of the last century. The isolationism of Washington was replaced by the imperialism of McKinley; but it was an imperialism with a bad conscience and of unbelievable crudity, to refer again to the tactics of Theodore Roosevelt not only in the Alaska boundary dispute but also in the Panama Canal negotiations. The crudity was perhaps best expressed in the phrase attributed to Representative Campbell, "Mr. President, what is the constitution betwixt friends?" Conscience reasserted itself in the reduction of tariffs on newsprint after the Reciprocity Treaty was defeated by Canada in 1911 and in the repeal of measures designed to improve the position of other powers especially Great Britain in the use of the Panama Canal. Rejection of the Reciprocity Treaty by Canada was a protest against crude imperialism as was to some extent the defeat of the Republican Party in the United States. The election of Wilson, the reluctance to become embroiled in the First World War, the lofty sentiments expressed by Wilson on the entry of the United States into the war, and the refusal to accept the League of Nations were evidence of an uneasiness about imperialistic tendencies.

Such uneasiness proved in itself, however, to be a spur to further imperialistic concern. Loans to European countries were interpreted as debts and consequently as subject to the payment of interest and ultimate repayment. In the words attributed to President Coolidge, "They hired the money, didn't they?" Insistence on recognition of debts strengthened the plea of debtors for loans from the United States with which interest on debts to the United

States could be paid. The burden of reparations on Germany was met by various devices in Germany and without, ranging from inflation to the expedients of the Young and the Dawes plans. This great merry-go-round began with President Harding's interest in normalcy and ended with President Hoover's earnest statement that the world was in a new financial era and that technological advance was such that it could support indefinite improvement in standards of living. Unhappily not even presidential assurances were sufficient to prevent the financial crash of 1929 and the consequent depression. The whole elaborate house of cards collapsed. Great Britain went off the gold standard, Hitler came into power and Franklin Roosevelt became President. Neither uneasy imperialism nor uneasy isolation had paid off.

Consequently the depression was marked by a return to isolationist and domestic policies. Roosevelt, without acknowledgment to Thoreau, proclaimed that the only fear we have to fear is fear. The United States was concerned with legislation designed to protect her from foreign entanglements. Isolationist policies had been evident in high tariffs especially the Hawley Smoot tariff and had compelled counter measures in other countries, notably the Ottawa agreements of the British Commonwealth. During this period of retreat Hitler began a programme of rapid expansion in Germany paralleled to some extent by a similar programme of Mussolini in Italy and by attacks on Manchuria from Japan. Great Britain became involved in a long series of manoeuvres ranging from the abdication of Edward VIII and the visit of the King and Queen, to the meetings in Munich, designed to delay the inevitable struggle, and to allow her to prepare with all possible energy during the delay, at the same time impressing on North America a reluctance to engage in war and a determination to become involved only on extreme provocation. The results scarcely need to be detailed since we are much too familiar with the history of the war and the phases leading to our present discontents.

Lessons had been learned in the First World War of which full advantage was taken in the Second World War. Systems of controls had been worked out during the long period of preparation after 1934 and were immediately applied on the outbreak of war. Devices elaborated in Canada were used by American propagandists as illustrations of the possibility of improvement in American controls with the result that Canadians reading the literature of American propagandists obtained a picture of their superior virtues. In the United States the dangers of large loans to allies were avoided by the ingenious system of lend lease. As a result of the applications of the lessons of the First World War the peace has been characterized by new developments. Fear of Germany in the east and the west following two world wars has prevented the signing of peace treaties and left that country divided between various interests. Fear of a depression during a possible reconversion period from war to peace, which followed the First World War until the system of American loans for repaying Americans debts was devised, has favoured an emphasis on military expedients ranging from the Marshall Plan to the Atlantic Pact by which full employment can be assured. Militarism becomes a necessity for the continued export of goods and for continued employment.

The emphasis on Communism has been an important element in persuading Americans that they must buy their own business. It would be unwise to comment on American foreign policy other than by reference to American writers. Archibald MacLeish in an article on "The Conquest of America" in the *Atlantic Monthly* (August 1949) wrote, "Never in the history of the world was one people as completely dominated intellectually and morally by another as the people of the United States by the people of Russia in the four years from 1946 through 1949. American foreign policy was a mirror image of Russian foreign policy. Whatever the Russians did, we did in reverse." H. Ickes in the *New Republic* (October 17, 1950) wrote, "We have been subjugated by Russia because of our fear of Russia." "I thank God that Roosevelt is not here now to see a greater and a stronger America not on its knees but on its hands and knees grovelling before dangers of its own imagining." The outsider can perhaps see more clearly than these writers the truth of their remarks in the work of the Committee on Un-American Activities, in the reign of terror introduced as a result of a revival of a system of informers in ex-Communists' rackets, in trials and penalties, and in rumours of suicides such as one heard in the stories from Germany and Italy. Bertrand Russell has described totalitarian countries as condemning people to lives of perpetual enthusiasm. In turn we seem to be condemned to lives of perpetual hate. Unity and coherence achieved in the United States by animosity against Great Britain such as was exploited by the Irish has necessitated animosity against Russia.

V

It has been largely in response to the pressure from American imperialism that Canada has developed her own type of imperialism. Nova Scotia entered Confederation on condition that the resources of the larger federal unit should be used to compel the United States to recognize her rights in the fisheries. Canada had no hesitation in using her influence to prevent a treaty between Newfoundland and the United States which seemed to threaten her bargaining position in the fisheries. The Act of Union was designed to enable Ontario and Quebec to develop transportation facilities which would meet American competition. Expansion of Confederation westward checked encroachments from the United States. The policy of the Dominion in the development of the prairie provinces was evident in the support of the Canadian Pacific Railway and in land policies designed to check American aggression. The character of Canadian imperialism became evident in the growing insistence on nationalism shown in the defeat of the Reciprocity Treaty, in the peace treaty of Versailles, in the Statute of Westminster, and finally in the acquisition of Newfoundland. It would not be difficult to collect a series of slogans comparable to those of the United States illustrating our imperialistic ambitions. Fittingly enough they might begin with the comment made at the beginning of the century, "The twentieth century is Canada's."

Pressure of American imperialism on Canada has been evident in a wide range of activities. Freedom has been perceptibly narrowed as a result of American hysteria. If a member of the academic staff of a Canadian institution wishes to take advantage of even a temporary appointment in the United States he must choose his relatives and his friends with much greater care than an American citizen. Presumably he should not belong to a party such as the C.C.F. or be involved in any discussions which might make him suspect as a threat to the American way of life. A Canadian citizen may not only be refused admission to the United States but the fact may be drawn forcibly to the attention of the public in American publications. Freedom of speech and of the press has not only been weakened directly as a result of American influence but indirectly as Canadians yield to the acceptance of standards imposed by the United States. The academic world will not overlook an attempt to humiliate its most brilliant scholars by American immigration officials nor tolerate affronts to its pride at a most sensitive point. In 1950, the middle of the twentieth century, a holy year, surely the lowest ebb in any civilization was reached when it was possible to threaten the lives of thousands of people with atomic bombs, with scarcely a protest in the interest of common humanity.[37] Fortunately we can still turn to Great Britain and Europe. Scholars turned back at the American border have felt much satisfaction at being given honorary degrees by British universities. But everyone must be disturbed by the appearance of the problem of the American refugee. The imposition of oaths for teachers has involved profound disturbances to American academic life and led to a concern of American scholars in appointments outside the United States.

The dangers of using militarism as a device for maintaining full employment which appear in American policy described as a mirror image of Russian policy are shown more sharply in the mirror image of Russian policy which we have in Canada. Ideologies are the fig-leaves of militarism. T. S. Eliot has said that "a true satellite culture is one which, for geographical and other reasons, has a permanent relation to a strong one,"[38] and to the reasons against consenting to its complete absorption into the stronger culture. The first: "it is the instinct of every living thing to persist in its own being"; the second, "the satellite exercises a considerable influence upon the stronger culture; and so plays a larger part in the world at large than it could in isolation." "The survival of the satellite culture is of very great value to the stronger culture."[39] He proceeds to suggest "that both class and region, by dividing the inhabitants of a country into two different kinds of groups lead to a conflict comparable to creativeness and progress"—a point emphasized almost two centuries ago by David Hume. "I do not approve the extermination of the enemy; the policy of exterminating or, as is barbarously said, liquidating enemies, is one of the most alarming developments of modern war and peace, from the point of view of those who desire the survival of culture. One needs the enemy. . . . The universality of irritation is the best assurance of peace."[40] "Just because and only because the natural spirit of conflict finds such frequent scope does human society hold together and on the whole present a pacific aspect." "The great majority could not live without oft-recurrent squabbles."[41]

I have ventured to digress in these remarks as a means of suggesting that criticism of the United States and of Canada is intended to be in the interests of both and to protest against a policy of American militarism which compels dependence on the United States. The distortions of the Canadian mirror may be more clearly seen by a description in greater detail of the process by which what is called light is reflected. Sixty percent of the circulation of periodicals is dominated by Americans, a reduction from the 80 percent of a couple of decades ago, but a reduction offset to an important extent by the influence of radio broadcasting to be supplemented shortly by television. The rapid advance of technology in the field of communication and the vast American market make it inevitable that the United States should dominate English culture and that it should exercise a powerful influence on French culture even though the latter is protected by language.

If the American nation has been described as "on its hands and knees grovelling before dangers of its own craven imagining," the Canadian people might be described as standing on their heads. The most significant indication was the size of the Liberal majority in the last election. No satisfactory explanation of this phenomenon based on the assumption that Canadians act rationally has been forthcoming. It has been argued that the Liberals showed themselves to be far more competent in handling election campaigns, that Mr. Drew alienated support by his application of provincial antics to the federal field, that elation over the retirement of William Lyon Mackenzie King spurred Liberals to a new pitch of enthusiasm and so on, but these arguments are not adequate and are scarcely sufficient to explain why the electorate felt that a strong opposition was not important. It may be suggested that militarism played its role in that emphasis was given to it by all parties and that such emphasis could have no other effect than strengthening the party in power.

Nothing is more ominous than the facility with which the tendency toward totalitarianism has enabled governments to create and exploit crises particularly in periods preceding elections. Mr. Churchill's genius as a politician in the British elections was evident in his recognition of this fact, shown in the popularity of his proposal for a discussion of the problem of the cold war at top levels. In Canada the threat of Communism was stressed by the Conservatives as a means of smearing the C.C.F. In turn the C.C.F. was compelled to stress its own reactionary characteristics in order to evade criticism. The weakness of smaller parties evident in their tactics became a source of strength to the Liberals. As a result the political shape of Canada began to assume characteristics similar to those of Russia. A politburo in Canada comparable to and paralleling that of Russia effectively diverts attention from its character by pointing to the dangers of the politburo in Russia. The distortion of Canadian political life has been evident in the attempts of the ambitious to acquire prestige by exploiting Russian stupidity. The stupidity of Russians inciting the attacks of ambitious Canadian leaders has been paralleled by the stupidity of Canadians in recognizing the incitement. In the field of labour the distortion has been evident in the hardening of labour organizations, following much-publicized purges of Communists, by a more rigid discipline, and by a greater capacity to exact demands and a great determination to enforce them.

In Canada a powerful bureaucracy, in part a product of bilingualism, built up in the depression and during the war, continued to exercise a powerful influence in a period of peace to an important extent by insisting that war had never ceased. Centralization which developed rapidly during the depression and was accompanied by a strong civil service and a decline of cruder forms of patronage was followed by the growth of provincial autonomy parties. The stupefying effects of the bureaucracy have been partly a result of the problem of a dual language in government and administration which blunts political edges. W. L. Mackenzie King as prime minister emphasized the importance of a French partner but his successor Mr. St. Laurent has no younger and outstanding individual English partner. He has rather a group of younger English members of the Cabinet anxious ultimately to assume his mantle. Mr.

King's talent for eliminating rivals at the appropriate time has been to some extent denied his successor.

Of more serious consequence has been the destruction of our sense of humour which has accompanied a lack of sense of proportion and a lack of criticism. No one can be a social scientist in Canada without a sense of humour. I offer this remark as a footnote to an understanding of Stephen Leacock. The appointment to office of the President of the Canadian National Railways because he had been a deputy governor of the Bank of Canada and had built up prestige in the Wartime Prices and Trade Board by violating the traditions of anonymity in the civil service has created no ripple of amusement throughout Canada. Within the space of a week or so he appeared as an authority on trade, banking, combines, and railways. In the words of Anita Loos: "A joke is a joke but no one wants to die laughing." The hazards of our profession are becoming serious.

The results of an overwhelming majority for the Liberal Party in the federal government and of their control of the Senate and the bench have been evident in various directions. This domination has left individual provinces as the only opposition, enabled the premier of a province to become the Conservative Leader of the Opposition, and accentuated the problem of federal government. Parties other than the Liberal Party tend to dominate the provinces. Consequently dominion-provincial relations occupy a more important role in Canadian politics. Development of opposition from labour and the C.C.F. in some provinces has been followed by coalitions of Liberals and Conservatives. General disequilibrium and instability have necessitated enhancement of the power of the Dominion, evident in the abolition of appeals to the Privy Council and in attempts to develop formulae for amendments to the constitution. The tendency towards centralization has accentuated an interest in defence and created an impasse strengthening the influence of the United States. The sense of omnipotence derived from an emphasis on the theory of the divine right of legislatures developed in the federal government compels a sense of ominipotence in provincial governments; it is no accident that the Province of Ontario outraged a sense of justice by retroactive legislation and that the federal government created a sense of futility by disregarding its own regulations in the Department of Justice in dealing with the Combines Report on Flour Milling.

The divine right of legislatures has contributed to the breakdown of the federal structure. Destruction of political relations between the parties of the federal government and those of many of the provinces has widened the gap between the provinces and the Dominion. A decline in the practice of the federal government of recruiting politicians from the provinces and a resort to that of building up the federal cabinet from federal politicians have sharpened the differences between the provinces and the Dominion. The problem has become more acute as a result of increased emphasis on central monetary policy. The basis of federalism in which the provinces maintained or acquired control over natural resources has been largely destroyed as a result of an increasing emphasis on monetary policy and particularly of large-scale resort to income taxes. Provinces and municipalities have been compelled to rely to an increasing extent on other taxes and the control of the federal government has been strengthened by division of powers and decline of the principle of taxation without representation.

Decline of the principle of taxation without representation has meant resort to agreements and large-scale arrangements for transfers between regions. Conflicts arising from the dependence of some regions on European markets and of other regions on American markets and the political power of the densely populated regions dependent on the United States compel resort to political patronage on a large scale to areas less effectively represented. Federal patronage has been essential to the prosperity of agriculture in western Canada.

The extreme complexity of government and the inability of the average citizen to understand its problems increases the responsibility of the bureaucracy. The latter are compelled to insist on democracy as a means of hiding the necessity of working contrary to democratic principles. In turn scepticism, as indicated in this chapter, of discussions of democracy is inevitable. The franchise has been extended and redistribution carried out with due regard to the advantages of the party in power, and large numbers have been appealed to by the parties concerned—all measures designed to strengthen democracy and calculated to work out to the advantage of the bureaucracy. The great art of political success dependent on keeping Scottish Presbyterians and French Canadians in the same party is no longer necessary.

It is impossible in this chapter to discuss exhaustively the effects of the enormous majority of the Liberal Party in Canadian life. Politics can no longer be discussed in terms of principles and with reference to abstractions. The power of the bureaucracy precludes an appeal to principles and compels concentration on details. Effective criticism becomes impossible with the deliberate attempt to focus attention on external affairs and the emphasis on the necessity of presenting a unified front to the point that essential control over military matters, regarded as the essence of sovereignty, is geared to the United States. There is still a fable to the effect that supping with certain mythological figures should only be done with a long spoon. We can appreciate the words of James Fitzjames Stephen about " 'Le self government', which not infrequently means the right to misgovern your immediate neighbours without being accountable for it to anyone wiser than yourself."[42]

It may be argued that all these problems will be solved by the abolition of appeals from the Supreme Court to the Privy Council. Dicey has remarked that "federalism substitutes litigation for legislation" and if we are to understand the prospects of success of the federal system we must pay some attention to the nature of the body before which litigation is carried out. The extent to which the new powers of an enlarged Supreme Court may be able to solve the problems of a federal state should engage the attention of citizens concerned with the continuation of the traditions of common law. Greater emphasis will be given to the written constitution particularly in problems of civil rights. Federal constitutions provide hiding places for vested interests.[43] The rights of property entrenched in written constitutions restrict possible developments of socialism such as have been evident in Great Britain. Sharp differences emerge between business and government. In federal constitutions emphasizing the traditions of Roman law in common law countries supreme courts occupy a crucial position. Common law traditions assume that the state is part of the law and the subject has greater difficulty in separating himself from the state. Change is consequently more gradual and less subject to revolution. Constitutions are largely protected from drastic revision. But Roman law tradition favoured by written constitutions in the United States and in members of the Commonwealth leans toward imperialism, and threatens the beneficial effects of common law in Western civilization. Without a recognition of the flexibility of common law the remark of Dean Pound that "legal precepts are almost certain to lag behind public opinion whenever the latter is active and growing" will become extremely pertinent. These fundamental problems face the Canadian courts and the Canadian people.

As a result of a firm belief in the impossibility of the spread of Communism in common law countries and in the danger of American imperialism in exploiting us through its propaganda about Communism I have felt compelled to seize this opportunity to describe our difficulties. The sense of terror which has seized on Canadian life has made it more imperative that I should regard the 150th anniversary of the University of New Brunswick as an occasion on which our faith in the traditions of common law, which were reflected after the American revolution in the founding of this university, could be reaffirmed.

Notes

1. James Bryce, *The Ancient Roman Empire and the British Empire in India: The Diffusion of Roman and English Law throughout the World: Two Historical Studies* (London,

1914); see also Lord Macmillan, *Two Ways of Thinking* (Cambridge, 1934). The Romans, like the British, presumably because of a strong oral tradition, made no distinction between constitutional and other laws.

2. "The King himself ought not to be subject to any man, but he ought to be subject to God and the law, since law makes the King. Therefore let the King render to the law what the law has rendered to the King, viz., dominion and power, for there is no King where will rules and not the law" (Bracton). Various writers have discussed the origins of common law in England to show that it consisted of customs which existed in unwritten form and that it was necessary to discover these customs through the use of the jury system and the calling together of representatives of different communities in Parliament. In the words of Pollard, "A foundation of common law was indispensable to a house of common politics." The terms writ, oath, witness, and possibly gallows did not originate in France. Parliament was concerned with the protection of individuals and not with the provision of privileges enabling members to abuse individuals outside its walls. Until the rebellion of the seventeenth century it was pre-eminently judicial rather than legislative. "The pleasure of the Prince has the force of law" as a maxim of Justinian's *Institutes* was recognized in England after Alfred and after the Norman Conquest, which made England one great fief in the hands of the king, facilitated the establishment of common law. See C. H. McIlwain, *The High Court of Parliament and Its Supremacy: An Historical Essay on the Boundaries between Legislation and Adjudication in England* (New Haven, 1934).

3. The weak point of common law has been stated to be its lack of executive power. Legislative amendments to correct the mistakes of the courts are apt to be unsatisfactory. Rome linked the army and the law and in the United States militarism became important to the executive power.

4. For a description of the veneration in which the common law was held under Eldon and the attacks of Bentham who regarded it as "a fathomless and boundless chaos made up of fiction, tautology, technicality and inconsistency, and the administrative part of it a system of exquisitely contrived chicanery, which maximizes delay and denial of justice," see J. F. Dillon, "Bentham's Influence in the Reforms of the Nineteenth Century," *Select Essays in Anglo-American Legal History* (Boston, 1907), pp. 494 ff.

5. Goldwin Smith, *Essays on Questions of the Day* (New York, 1893), p. 98.

6. M. J. Bonn, *Wandering Scholar* (New York, 1948), p. 89; see also George Dangerfield, *The Strange Death of Liberal England* (New York, 1935). It was in reply to Isaacs on the difficulties of Carson in the Ulster problem that Spender remarked, "You are all lawyers and lawyers never put one another in prison" (Harold Spender, *The Fire of Life* [London, n.d.], p. 150).

7. Bonn, *Wandering Scholar*, p. 101.

8. "Anglo-Saxon societies prefer to regulate a maximum of conduct by convention and are inclined to believe that only behaviour regulated by folkways and mores is free. By contrast, societies influenced by the tradition of Roman law lean toward statutory regulation, where definitions are clear cut and the sources of pressure are explicit and visibly organized" (Karl Mannheim, *Freedom, Power and Democratic Planning* [New York, 1950], p. 51).

9. Sir Henry Taylor, *Notes from Life—The Statesman* (London, 1878), pp. 384-85.

10. Cited by E. S. Corwin, *The Twilight of the Supreme Court* (New Haven, 1934), p. 207.

11. As early as 1824 Macaulay could write: "Our legislators, our candidates, on great occasions even our advocates, address themselves less to the audience than to the reporters. They think less of the few hearers than of innumerable readers." *On the Athenean Orators*. Sir Austin Chamberlain referred to the practice which had developed of ministers reading from type-written manuscripts (*Politics from the Inside* [London, 1936], p. 245).

12. Frederic Harrison, *Autobiographic Memories* (London, 1911), I, 149. See Adam Smith, *The Wealth of Nations* (New York, 1937), p. 680.

13. I am indebted to Mr. F. M. Covert, Q.C., and Mr. G. Demerais, Q.C., for general views on this subject.

14. Max Lerner, *The Mind and Faith of Justice Holmes* (Boston, 1943), p. 72. See also *The Life and Labours of Albany Fonblanque*, ed. by his nephew, E. B. de Fonblanque (London, 1874), pp. 340-41.

> It was but the other day, however, that a most tender and touching sight was presented in Lord Carlisle's Court of Inquiry—Mr. Serjeant Wilkins weeping for Mr. Ramshay, his learned bewigged head bent to the table "like a lily borne down by the hail." Perhaps, prosaically, it was more like a cauliflower on a block, but let that pass. What we have to consider is the zeal, or the fee-compelling-force, which can bow a wigged head to the table, and make the eyes overflow with tears such as either genuine pity, or genuine onion, elicits—tears such as learned serjeants shed. The eye that so weeps, however, must have seen a fee. An unfeed eye would on a similar occasion be as unmoved as a stone. The fee and the feelings go together; the word feeling, in legal diction, being derived from fee. What the precise charge for weeping is we do not pretend to know; nor whether it is set down in the brief as an extra, like consultation, or a refresher; but of late years we have had several exhibitions of this black grace. Chitty wept for Thurtell, and Fitzroy Kelly for Tawell, and lastly Wilkins for Ramshay. Sweet sensibility! says the tender-hearted reader; but how is it that this same sensibility of the learned is so capricious, and that the same wigged man, who blubbers over one client so affectingly, will throw another overboard without a hesitation or a scruple, Why make fish of one and flesh of another? Why so strain the duty of advocate and client in some show cases, and loosen it in others, as we see in this example?
> The complaisant husband who had napped during Caesar's visits, on finding that the same somnolency was expected from him by another gallant said, "I do not slumber for everybody." Mr. Serjeant Wilkins does not sob for everybody; but in common fairness and honesty he is bound to explain the rules of his service or disservice to his clients, specifying for which of them he goes through thick and thin, and which he throws overboard." (1851)

Timothy Healy remarked concerning opposing counsel who wept for his client that it was the greatest miracle since Aaron struck the rock. The high cost of law has an advantage in checking litigiousness but emphasizes justice for those who can afford to pay. Lack of protection to the economically weak in England probably hastened the growth of capitalism.

15. Legal business in Quebec is divided between lawyers and notaries, lawyers being primarily concerned with litigation. In Anglo-Saxon provinces most of the legal

business is handled outside the courts. The harshness of common law has favoured a steady encroachment by elements of civil law through admiralty law, international law, and the growth of statutes. Pressure of the courts leads to the enactment of statutes. Workmen's Compensation legislation relieved the courts of enormous numbers of employer-employee suits.

Bentham carried the demands for written law to great lengths. "So long as there remains any of the smallest scrap of unwritten law unextirpated, it suffices to taint with its own corruption—its own inbred and incurable corruption—whatsoever portion of statute law has ever been or can ever be applied to it." On the other hand Ehrlich stated, "It is a fair question to ask whether codification of the law may not be objectionable on this ground alone, viz., that it enforces on human life the will of the state in a thousand instances, although frequently the state is not interested in the least that such should be the case." Codes and statutes impose a heavy burden on language. "It is accordingly the duty of a draftsman of these authoritative texts to try to imagine every possible combination of circumstances to which his words might apply and every conceivable misinterpretation that might be put on them and to take precautions accordingly" (J. F. Stephens). Changes in languages necessitate the constant attention of the courts. "A word is not a crystal, transparent and unchanged; it is the skin of a living thought and may vary greatly in the colour and content according to the circumstances and the times in which it is used" (Holmes). Lawyers tend to favour words of obvious meaning which will remain static over a long period. The common law tends to mould facts to suit words and Roman law to mould facts to suit writing. Civil law proceeds from principles to customs, common law from customs to principle; civil law searches for principle, common law for precedent. The precision of the French language favoured the growth of a code while the lack of precision of the English language favoured an appeal to precedents. See Sir Ernest Gower, *Plain Words* (London, 1948).

16. Valentine Williams, *The World of Action* (Cambridge, 1938), p. 309.
17. Corwin, *The Twilight of the Supreme Court*, p. 54.
18. Lerner, *The Mind and Faith of Justice Holmes*, p. 83.
19. Ibid., pp. 85–86.
20. See F. W. Maitland, *English Law and the Renaissance* (Cambridge, 1901). Common law with common politics and parliament probably checked the influence of religion and facilitated the break with Rome. See also W. Stubbs, "The History of the Canon Law in England" in *Select Essays in Anglo-American Legal History* (Boston, 1907), pp. 252 ff. The Inns of Court emphasized disputations in the oral tradition and defeated the threat of civil law. Following the spread of printing the position of the common law was secure.
21. Brooks Adams, *The Emancipation of Massachusetts, the Dream and the Reality* (Boston, 1919). The conflict between Puritans and Cavaliers in England was to some extent transferred to North America in the settlement of Puritans in New England and of Cavaliers in Virginia and in the war between the states. The migration of Puritans to New England before the Civil War was followed by an intensified bigotry and an intolerance which persisted in American life and which reinforced the Roman law elements of the American constitution. See P. D. Reinach, "The English

Common Law in the Early American Colonies" in *Select Essays in Anglo-American Legal History*, pp. 367 ff. See also Frederick Pollock, *The Genius of the Common Law* (New York, 1912), pp. 56 ff. New England colonies vested with legislative power emphasized the scriptures and developed systems varying with the common law and a disrespect for lawyers, and weakened the position of the courts.

22. Cited in Champ Clark, *My Quarter-Century of American Politics* (New York, 1920), II, 130. The printing of laws in the colonies meant the establishment of printing plants, excess capacity, the development of newspapers, and the emergence of postmasters.

23. Ibid., II, 133.

> But the lawyers in any community are often the most intelligent men in it. They have the broadest minds and usually the narrowest sympathies. Their intelligence, their knowledge of human nature, especially in its weaknesses, their intimate touch with almost all its affairs, and their acquired power of statement have given them the preponderant influence in affairs of state which they traditionally enjoy in America; while their narrow sympathies, their addiction to technicalities, their subserviency to vested wrongs as well as vested rights, and their tendency to unscrupulousness in methods and to scepticism as to the good in humanity as a whole, have made that influence a barrier to progress. (Herbert Quick, *One Man's Life* [Indianapolis, 1925], pp. 330–31)

24. Corwin, *The Twilight of the Supreme Court*, pp. 180 ff.

25. Brooks Adams, *The Theory of Social Revolutions* (New York, 1913), p. 97.

26. Harriet Monroe, *A Poet's Life: Seventy Years in a Changing World* (New York, 1938), p. 40.

27. Irving Stone, *Clarence Darrow for the Defense* (New York, 1941), p. 369.

28. Ibid., p. 368.

29. J. Alsop and T. Catledge, *The 168 Days* (New York, 1938), p. 3.

30. Ibid., p. 135.

31. Arthur Cheney Train, *My Day in Court* (New York, 1939) p. 368.

32. Ibid., p. 361.

33. Ibid., p. 73.

34. Stone, *Clarence Darrow for the Defense*, p. 72.

35. Ibid., pp. 164, 515.

36. Ibid., p. 107. With this influence of the lawyer on public opinion must be contrasted the tragedy of lynching. From 1889–1932, 3,753 lynchings, white and black, were recorded. J. H. Chadbourn, *Lynching and the Law* (Chapel Hill, 1933). For an illustration of casual interest in the problem see *The Intimate Note-books of George Jean Nathan* (New York, 1932).

37. "The ancient rule of *hosti etiam fides servanda* is ended" (Alfred Vagts, *A History of Militarism* [New York, 1937], p. 431).

38. T. S. Eliot, *Notes towards the Definition of Culture* (London, 1949), p. 54.

39. Ibid., p. 55.

40. Ibid., p. 59.

41. George Gissing, *The Private Papers of Henry Ryecroft* (London, 1914), pp. 92–93.

42. James Fitzjames Stephen, *Liberty, Equality, Fraternity* (London, 1874), p. 268.

43. "To support vested interests is what lawyers are paid for and what courts are made for" (Brooks Adams, *The Emancipation of Massachusetts*, p. 130).

CHAPTER FOUR

The Press, a Neglected Factor in the Economic History of the Twentieth Century

May I begin by expressing my appreciation of the honour which has been shown me by asking me to deliver a lecture in memory of the late Josiah Charles Stamp, First Baron Stamp of Shortlands. I am particularly happy because it enables me to pay a tribute to his work in Canada. The late Viscount Bennett, sensitive to the prestige of the name of Stamp, asked him to investigate the problems of marketing grain in Canada and the results were presented in the Stamp Report, an important document in the history of marketing.

Under the terms of the Trust deed this lecture must have as its subject "the application of economics or statistics to a practical problem or problems of general interest" and the subject must be treated "from a scientific and not from a party political standpoint." In these days of totalitarian tendencies it might be argued that these clauses involve a contradiction, or that they are a directive requiring consideration of the problems of totalitarian states. Knowing that the latter alternative might well be regarded as sacrilege to the memory of the man in whose name the Trust was created, I am compelled to consider the problem of possible contradiction. In the search for an answer to this problem I have been fortunate in finding in the works of Graham Wallas, who exercised such an important influence on the thought of Lord Stamp, a possible clue. It may seem that I feel directed to lecture on the subject of the subject or directly on the conditions which have led to the formulation of these precise conditions, and that in this way I have reconciled the contradiction, and indeed I hope that this will be partly true. Thorold Rogers has remarked that among the calamities afflicting political economy is the fact that "all or nearly all its fallacies are partially true,"[1] which remark possibly provides a basis for his comment that "a cheap investment [is] to be made in popular delusions. I know no safer speculation."[2] I hope that my discussion of popular delusions will not prove an unsafe speculation to my audience.

I am aware that I am only presenting a footnote on the work of Graham Wallas, but it should be said that the subject of his work was in itself inherently neglected. He chose in his later publications to concentrate on the problem of efficiency in creative thought. He emphasized the importance of the oral tradition in an age when the overpowering influence of mechanized communication makes it difficult even to recognize such a tradition. Indeed the role of the oral tradition can be studied only through an appraisal of the mechanized tradition, for which the material is all too abundant. The lecture, one of the last vestiges of the oral tradition, has been overwhelmed by the written tradition and the examination system in spite of the noble efforts to support its continuance by foundations such as that in which I speak. And even such lectures as these are destined for print.

I have attempted elsewhere to develop the thesis that civilization has been dominated at different stages by various media of communication such as clay, papyrus, parchment, and paper produced first from rags and then from wood. Each medium has its significance for the type of script, and in turn for the type of monopoly of knowledge which will be built and which will destroy the conditions suited to creative thought and be displaced by a new medium with its peculiar type of monopoly of knowledge. In this lecture I propose to concentrate on the period in which industrialization of the means of communication has become dominant through the manufacture of newsprint from wood and through the manufacture of the newspaper by the linotype and the fast press. Physics and chemistry have been largely concerned, notably in the study of electricity with its possibilities for increased speed of communication. The telegraph and particularly the telephone[3] were significant to the expansion of journalism.

The conservative power of monopolies of knowledge compels the development of technological revolutions in the media of communication in marginal areas. In the first half of the nineteenth century in Great Britain "taxes on knowledge," as they were called, ensured a monopoly position for *The Times* such that restriction in the use of paper for newspapers favoured an increase in the production of periodicals and books. This increase was enhanced by an expanding American market unprotected by copyright legislation. American writers were driven into the field of journalism in the United States, particularly as a free press was protected under the Bill of Rights. Removal of taxes on knowledge about the middle of the nineteenth century in

Great Britain favoured the importation of improvements in techniques in the production of newspapers from the United States and the growth of the new journalism in Great Britain and on the Continent in the latter part of the nineteenth century and in the twentieth century, notably in the Boer War and in the First World War.

It is necessary at this stage to indicate briefly the major technological changes in communication in North America and in turn in Great Britain and to a lesser extent in Europe. As a result of the use of wood in the manufacture of newsprint the price of the latter declined from 8 1/2 cents per lb. in 1875 to 1 1/2 cents in 1897. With decline in price larger quantities were used and new inventions were developed at later stages in the production of the newspaper to eliminate a series of bottlenecks. Introduction of the linotype in 1886 was followed by a reduction in the cost of composition by one-half. The demands of type-setting machines for large quantities of legible material compelled the use of typewriters.[4] The cost of printing was vastly reduced with more efficient presses. A double supplement press installed by the New York *Herald* with a capacity of 24,000 copies of 12 pages each per hour was far surpassed by a quadruple press installed by the New York *World* in 1887 with a capacity of 48,000 copies of 8 pages per hour and in turn by an octuple press with a capacity of 96,000 copies of 8 pages per hour in 1893. Improved methods of producing printed paper were supplemented by methods for reproducing illustrations; zinco and the half-tone facilitated reproductions of photographs after 1880. Pulitzer's use of the cartoon had contributed to a quadrupling of circulation by the end of the first year. His success in increasing circulation with pictures was immediately followed by others. The multi-colour rotary press was introduced in the early nineties with pictures. By 1900 nearly all daily papers in the United States were illustrated.

With the turn of the century in the United States, a marked increase in the size and circulation of newspapers was accompanied by increasing costs of newsprint and attempts on the part of newsprint companies to strengthen their position by amalgamation, notably in the formation in 1898 of the International Paper Company, composed of nineteen companies. The newspapers began to organize in opposition to a threat to increase prices. With the enormous advantage of control over publicity, they exercised sufficient political pressure to secure the reduction and abolition of tariffs on mechanical pulp and newsprint from Canada. Sensitive to the influence of newspapers over public opinion, Theodore Roosevelt launched a conservation campaign with the slogan "We are out of pulpwood." His successor, W. H. Taft, from all the controversy over the Reciprocity Treaty in 1911 emerged with the definite result of a low tariff on newsprint. Under Woodrow Wilson, the Democratic president, restrictions were removed on imports of newsprint. The success of the efforts of American newspapers was evident in a price of newsprint of 2 cents per lb. or $40 per ton in 1914.

The Canadian provinces with their control over large areas of Crown land followed policies designed to compel the importation of American capital and the establishment of paper-mills in Canada. Ontario imposed an embargo on pulpwood cut on her Crown lands and was eventually followed by Quebec in 1910 and New Brunswick in 1911. Large rivers provided cheap navigation and large power sites, and with proximity to the chief American markets by rail and favourable differential railway rates, governmental encouragement, and low labour costs, the province of Quebec was particularly successful in securing the construction of newsprint mills. Since four tons of raw material were necessary to produce a ton of newsprint, transportation rates were an important factor in the choice of sites.

During the war period prices increased. Newsprint cost $69 a ton in 1918 and after a sharp increase to $130 in 1920 attained a level of $75 per ton in 1922. With the stimulus of higher prices newsprint mills were established on a large scale. The annual capacity of Canadian mills increased from 715,000 tons in 1917 to 3,898,000 tons in 1930, doubling in the period from 1926 to 1930. Canada surpassed the United States as a producer of newsprint. Mills installed at later dates had the advantage of incorporating new inventions. In 1921 a paper machine 166 inches in width, running 1,031 feet per minute, established a record, but in 1927 machines of 270 inches were being installed. The greater efficiency of plants installed at later dates and the length of time involved in bringing larger plants into production led to a lowering of prices in the late twenties and to a sharp decline during the depression to $53 in 1931 and $46 in 1932. Competition of later more efficient plants with older plants became acute, with the result that an elaborate bond structure, developed during the period of rapid construction, was subjected to an intensive programme of reorganization accompanied by large-scale mergers.

In these drastic reorganizations hydro-electric power assumed a more important position. A newsprint mill includes plants for the production of mechanical pulp, sulphite pulp, and newsprint. A ton capacity of newsprint assumes roughly an installation of 100 horsepower of which 85 percent is used for mechanical pulp or groundwood. Attempts to increase the proportion of the cheaper mechanical pulp above 75 percent of the total content of newsprint have assumed increasing dependence on hydro-electric power. Since power sites are related to geographic considerations such as geology, topography, size of lakes (which serve as storage basins in a relatively severe cold season), size of rivers, and rainfall, and since they involve an enormous initial capital investment, operation at capacity will become an important factor in determining the size of paper mills. Since prices of newsprint tend to be held down by the strong position of newspapers, attempts will be made to divert hydro-electric power to municipal and industrial purposes. Large metropolitan papers such as the *New York Times* and the *Chicago Tribune* have attempted to strengthen their position by assuming direct control over newsprint mills,[5] but the possibility of adapting power sites and newsprint mills to the demands of such newspapers is limited. Yet mills which are not controlled by newspaper companies have a narrower market in which to sell their products. High prices of paper during the war and the post-war period were accompanied by numerous amalgamations of newspapers and the introduction of such high-speed equipment as steel cylinders, roller bearings, and ink pumps. Large newspaper owners such as Hearst and Scripps Howard, capable of providing a large steady market, are in a position to play off one newsprint mill against another and secure lower prices. Small newspapers in a weaker position have been linked in various chains to the same purpose. An attempt of the International Paper Company to combat these trends by acquiring newspapers in order to provide a more profitable outlet was defeated by the usual appeal to the importance of a free press and to the danger of control of power interests over the press. Insistence on freedom of the press became a powerful factor in the defeat of newsprint producers. As a result of the strong position of newspapers, newsprint companies have favoured reliance on the sale of hydro-electric power to municipal and industrial consumers, and paper has tended to become a by-product of power production, with the weak marketing position which characterizes by-products.

The increasing production of relatively low priced newsprint has been accompanied by an increase in *per capita* consumption of paper in the United States from 25 lb. in 1909 to 41 lb. in 1920 and 59 lb. in 1930. In 1914, 2,580 daily newspapers had a daily circulation of 41,131,611. Although the number of Sunday newspapers was only slightly reduced in the same period, their circulation increased from 16,479,943 to 32,371,092.[6] In the half-century from 1880 to 1930, in which the major technological changes which we have described took place and after which the radio becomes a more important competitor for advertising, the newspaper's dependence on advertising revenues increased from 44 percent to 74 percent. In twenty-three of the largest American cities advertising lineage increased from 662 million in 1914 to 1,293 million in 1929, with the most pronounced increase two years after the war, and declined to 746 million in 1933. The increase in percentage of revenue from advertising is misleading, since the reduction in the price of newspapers to one cent or one half-penny in 1900 and the relative rigidity of prices of newspapers is essentially designed to increase circulation and to attract advertising. It is safer to say with George Seldes that "the real publishers are the advertisers since their financial support of a publication is in most cases all that keeps it alive."[7]

The low prices of newspapers incidental to the need for circulation demanded by advertisers meant an emphasis on changes in the content of the newspaper which would attract the largest number of purchasers. The newspaper was made responsive to the market. The business office occupied a dominant position.[8] News became a commodity and was sold in competition like any other commodity.[9] Consequently it was classified in relation to the markets to which it was supplied and in relation to regions which produced it. In the words of Mr. Dooley, "Sin is news and news is sin." Charles Merz wrote that "it is doubtful whether anything really unifies the country like its murders."[10] In England R. D. Blumenfeld held that all grades of society were "more interested in crime mystery—particularly the murder of a woman—than in any other topic."[11] "There is more joy in Fleet Street over one lover who cuts his sweetheart's throat than over nine hundred and ninety-nine men who lived happily ever after" (A. P. Herbert). As a result of time and other considerations, various centres are compelled to specialize in the production and sale of news. News follows the sun and with the increasing importance of afternoon and evening papers generally happens in the afternoon. Chicago[12] as a great inland centre and with a definite time-lag from New York has been compelled to concentrate on crime. Mr. Dooley, commenting on Wilbur F. Storey, editor of the Chicago *Times* in the eighties, wrote: "They wanted crime an' he give it to them. If they wasn't a hangin' on th' front page some little lad iv a rayporther'd lose his job. They was murdher an' arson till ye cudden't rest, robbery an' burglary f'r page afther page, with anny quantity iv scandal f'r th' woman's page, an' a fair assortmen' iv larceny an' assault an' batthry f'r th' little wans."[13]

The malevolent influence and power of publishers has possibly been exaggerated, but, in the opinion of Seldes, Hearst proved "that news is largely a matter of what one man wants the people to know and feel and think."[14] Pulitzer was said to have "rather liked the idea of a war—not a big war—but one that would rouse interest and give him a chance to gauge the reflex in his circulation figures."[15] The publisher became concerned to secure consumer satisfaction: "never write to please the writer, write to please the reader."[16] In the words of Brisbane, the Hearst columnist, "nobody wants to know what *you* think. People want to know what *they* think."[17] Beaverbrook advised that "you must be ready to put into it your whole heart and soul, your stomach, your liver, your whole anatomy which will appear most of the time to be dangerously stimulating and on occasion positively revolting."[18] J. G. Bennett, Jr., stated that a journalist should be "inquisitive, catty, human, eccentric, generous and pernicious in turn, kindly and inexpressibly brutal from moment to moment, broadminded, well-read and suspicious."[19] A later American publisher insisted on printing "what any human being would be interested in—something that will not cause people to think, that will not even invite them to think—to enable them to forget rights and wrongs, ambitions and disappointments." "Reading next to sleeping is the best way to rest the mind" is Mr. Dooley's summary.

Arnold Bennett has written a pertinent description of the editor:

> to devise the contents of an issue, to plan them, to balance them; to sail with this wind and tack against that; to keep a sensitive cool finger on the faintly beating pulse of the terribly many-headed patron; to walk in a straight line through a forest black as midnight; to guess the riddle of the circulation-book week by week; to know by instinct why Smiths sent in a repeat-order, or why Simpkins' was ten quires less; to keep one eye on the majestic march of the world, and the other on the vagaries of a bazaar-reporter who has forgotten the law of libel; these things, and seventy-seven others, are the real journalism.[20]

News must be selected, its position on the page determined, the proper size of type chosen. "Nothing . . . possesses quite such power over people who like to believe they do their own thinking as that which seemingly leaves all thinking to them."[21] In perfecting "the art of lending to people and events intrinsically dull an interest which does not properly belong to them,"[22] the definition of journalism given by Arnold Bennett, journalists are exposed to unique advice from editors. The managing editor of the *Detroit News* held that "Four things were necessary to learn to write, the *Bible, Shakespeare,* the *Saturday Evening Post,* and the *Detroit News,*"[23] advice paralleled by that of Sir William Crawford that "copywriters must read the Bible, Kipling, Stevenson, and Burns because they know how to touch the human heart."[24]

Under the pressure of publishers and advertisers the journalist has been compelled to seek the striking rather than the fitting phrase, to emphasize crises rather than developmental trends. In the words of Escott, the journalist "increasingly seems to think that his duty to his paper requires the discovery of a new crisis or a new era." "The journalist has long been and will always remain, a stormy petrel, a fisher in troubled waters, one whose activities tend to excite, not to moderate, the popular passions."[25] Success in the industrialized newspaper depends on constant repetition, inconspicuous infiltration, increasing appeal to the subconscious mind, and the employment of tactics of attrition in moulding public opinion.[26] Northcliffe warned, "Remember the power of persistency in journalism."[27]

Journalism has been criticized on a wide front, particularly in England. Lord Salisbury described it as "an intelligent anticipation of events that never occur" and Leslie Stephen as "writing for pay upon matters of which you are ignorant";[28] "to be on the right side is an irrelevant question in journalism."[29] Humbert Wolfe (*The Uncelestial City*, Bk. 1, "Over the Fire" [New York; Alfred A. Knopf]) wrote

> You cannot hope
> to bribe or twist,
> Thank God, the
> British journalist.
> Considering what
> the man will do,
> Unbribed, there's
> no occasion to.

An American writer has been more savage: "The business of a New York journalist is to distort the truth, to lie outright, to pervert, to vilify, to fawn at the foot of Mammon, and to sell his country and his race for his daily bread.... We are intellectual prostitutes."[30] A more sober view states that "Journalism can never be history; its unceasing activities deprive it of the advantages of scientific inquiry. It cannot ever be the rounded truth, since the necessity of prompt presentation of what seems to be fact renders impossible the gathering and weighing of all evidence which bears upon the event which must be chronicled."[31]

Yet the journalist himself is not unaware of the pressure of the publisher and the advertiser as his castigations have made clear. Writers as the more sensitive and restive members of the community have been extremely sensitive to repression, particularly as they have been slow to organize resistance. Frank Munsey, who systematically bought, closed down, and amalgamated newspapers and was described as "one of the ablest retail grocers that ever edited a New York newspaper"[32] drew the following obituary notice from William Allen White: "Frank Munsey, the great publisher, is dead. Frank Munsey contributed to the journalism of his day the great talent of a meat packer, the morals of a money changer, and the manners of an undertaker. He and his kind have about succeeded in transforming a once noble profession into an eight percent security. May he rest in trust."[33] Philip Gibbs wrote regarding Northcliffe that he had never heard him "utter one serious commentary on life, or any word approaching nobility of thought, or any hint of some deep purpose."[34]

The character of news reflecting the demands of the advertisers particularly after 1900 emphasized discontinuity. Cazamian[35] suggests that the psychological results were evident in the tremendous success of moving pictures, which at its central roots sprang from the methods of discontinuity. Development of photography and of the cinema was paralleled by the increase in the use of illustrations in the newspapers particularly during the First World War. Established newspapers had reached new levels of pomposity illustrated by Pulitzer's comment that "The *World* should be more powerful than the President. He is fettered by partisanship and politics and has only a four years' term."[36] Their monopoly position was marked by conservatism in recognizing technological advance. The possibility of tapping lower levels of income and larger numbers for advertisers and recognition of the loosening of rules and habits during the war favoured the establishment of the tabloids.[37] Pictures[38] spoke a universal language which required no teaching for their comprehension. "The boob no longer believes anything he reads in the papers but he does believe everything he sees."[39] In the search for a wider circle of readers it became necessary to rely on topics with a universal appeal, notably on sex. Patterson of the New York *Daily News*, a most successful tabloid, introduced a new order for news—first love and sex, then money, murder, and health.[40]

Gauvreau, editor of an unsuccessful tabloid, hemmed by Patterson "into a pocket from which we could never fight our way out to our first daily million,"[41] stated that "no paper of mass appeal could afford to be without a staff astrologist or a palmist who could tell you how to improve your fortune."[42] "The space we devote to politics is a dead loss in circulation."[43] He wrote, "Never print anything that a scrubwoman in a skyscraper cannot understand,"[44] a caution paralleled by R. D. Blumenfeld's in England "never to forget the cabman's wife." We can appreciate his remark that he regarded himself as "the parasite clinging to the vitals of the kept press."[45] This audience will not be familiar with references to the excitement over "Peaches" Browning or over the Hall Mills murder case but they will be aware of the influence of the tabloids on the older press in a reference to Lindbergh. The Lindbergh flight, "the greatest torrent of mass emotion ever witnessed in human history,"[46] has been held to have changed the attitude of France towards the United States. In accounts of the kidnapping of the Lindbergh baby the press "pulled out all stops in its last great orgy of concentration on an individual case."[47] In the intense competition for circulation and for advertising a success was won by the use of reading matter and picture appeal in competition with the magazines and by the use of features which emphasized gossip about movie stars. Failure followed the refusal of advertisers to support a number of tabloids appealing to low-income groups. This could be cited against the statement of H. L. Mencken that no one ever went broke underestimating the taste of the American public.

The problem of adapting news to the needs of wider circulation led to greater dependence on feature material. The growing efficiency of press associations brought a decline in the number of scoops[48] claimed by individual papers and led to increasing dependence on local news and on features. Advertising demanded more space and larger newspapers, and to preserve a reasonable proportion of reading material it became necessary to depend on syndicates[49] for feature material. Newspapers formed their own syndicates for the creation of feature material or for the creation of news. Features were designed to secure a firm footing for newspapers in the home and to enhance the value of the paper from an advertiser's point of view. The large Sunday paper emerged in response to this demand and with it an insatiable appetite for features designed to appeal particularly to women and to children. The serial comic was admirably suited to this purpose. "The power of the popular strip over circulation is notorious." "Comics are the lifeblood of the entire syndicate business." Comic strip writers are compelled to consider the welfare of their characters with great seriousness if they wish to avoid a flood of protests from readers. Women were the object of attention in the features and in the news because of their influence on the purchase of commodities. Northcliffe[50] advised, "Always have one women's story at the top of all the main news pages of your paper."[51] "Women are the holders of the domestic purse strings. Men buy what women tell them." The demand for circulation by advertisers was a demand for entertainment and for a wide variety.[52] In the twenties the public turned from horse racing and liquor to column reading as a dissipation.[53] Newspapers carried the writings of columnists with wide and divergent points of view.

The effect of these changes was evident in the decline of the editorial as an influence on public opinion. Mr. Dooley said that the mission of the newspaper was "to comfort the afflicted and afflict the comfortable." The Hearst press[54] and, to a less extent, the Scripps Howard[55] press systematically and profitably exploited exploitation and newspapers were compelled to turn to other interests. As early as 1899 a study of newspapers showed that news of crime and vice, illustrations, and help wanted and medical advertisements, occupied an almost steadily increasing percentage of space, with an increase in circulation, while the opposite was the case with political news, editorials, letters and exchanges, and political advertisments.[56] H. L. Mencken has pointed out how the editorial writer lost prestige as compared with the news writer since less ability was required to express opinions than to present readable news and such ability was not as well paid. He writes: "I know of no subject in truth, save baseball, which the average newspaper even in larger cities discusses with unfailing sense and understanding." Another American writer states: "I doubt if there is an editorial page in America that is read by five per cent of the paper's readers. On the basis of results the average editorial page is the most expensive in the paper—in white space, composition and editorial labour."[57] A study of the American press concludes that the "modern commercial newspaper has little direct influence on the opinions of its readers on public questions. It probably seeks to reflect, not make opinions."[58] W. J. Bryan stated that "newspapers watch the way people are going and run around the corner to get in front of them."[59] Large circulations prevented the newspapers, as someone put it, from attacking anything but the man-eating shark.[60] You may remember in Samuel Butler's *The Way of All Flesh* the efforts of Townley to get Ernest Pontifex out of trouble, "to see the reporters and keep the case out of the newspapers. He was successful as regards all the higher-class papers. There was only one journal, and that of the lowest class, which was incorruptible." "The power of the press is the suppress."

In emphasizing the necessity of increasing the sale of newspapers, headlines and news have dominated the front page. With few exceptions advertising is confined to the later pages. It may be argued that "for practical purposes matter that is more directly profitable to the individual than to the community is called advertising and matter that benefits the community rather than the individual is called news," or that, in the words of Ivy C. Lee, "news is that which people are willing to pay to have brought to their attention; while advertising is that which the advertiser himself must pay to get to the people's attention."[61] The difficulties of separating the indirect sale of advertising, by making the front page sell the newspaper, and the direct sale of advertising has led to the rapid development of publicity men who have become skilled in disguising advertising material and planting it in unexpected places to be picked up as news.[62] While newspapermen have become suspicious of these news services and reluctant to introduce anything strange in their columns, publicity men have become more ingenious.[63] The politician, following the decline of the editorial page, has been particularly active in attempting to make the front page. Sam Chamberlain, an experienced newspaper man, wrote: "Give me the right hand column first page and I won't care what they put on the editorial page,"[64] and his views were echoed by La Follette in 1924: "I don't care what the newspapers print in their editorials about me if I can keep in their news columns."[65] The politician has actually begun to welcome abuse in the news. Frank Kent of the Baltimore *Sun* explained the power of the political boss in large cities: "the more violently he is denounced by the press the stronger the trend toward him."[66] George Seldes wrote, "I believe that Hearst as an ally of any politician is a form of political suicide."[67] With the radio, however, the politician capitalized on the limitations of the press, Franklin Roosevelt claiming that "nothing would help him more than to have it known that the newspapers were against him." Radio tipped the scales in favour of the individual candidate and overcame the influence of tons of Republican written propaganda. "There has been a landslide in every national election since the use of radio became general."[68] The success of Roosevelt followed the decline of the press as a medium of political expression.

Dependence on the front page has left its stamp on American political life in the character of personalities and of legislation. Keynoting which "implies the ability to . . . give the impression of passionately and torrentially moving onward and upward while warily standing still" (Lowry)[69] has been only one of the results. Bryce in *The American Commonwealth* has emphasized the significance of the absence of foreign problems in the United States in the nineteenth century (1891). To an important extent ambassadorial posts were a part of political patronage distributed among newspaper publishers. Representatives of the press who were distrusted in domestic appointments in the United States were acclaimed as ambassadors in other countries where individuals with such training were largely excluded from diplomatic posts. The results have been evident in a lack of experience and continuity in foreign problems, in the refusal of the United States to join the League of Nations, and in its blustering foreign policy since the establishment of the United Nations. "Agitation carried on by the press unsettles the public mind. The uncertainty prevents unity of action and that lack of unity of action prevents stability."[70]

Whitelaw Reid[71] remarked in the late nineteenth century that "news brings circulation and advertisements," but this view has been modified by the insistence that "there is no substitute for circulation."[72] Lasswell has commented that "literacy and the press are offsprings of the machine age. The press lives by advertising, advertising follows circulation and circulation depends on excitement."[73] In building up circulation itself and creating goodwill the newspaper attempts to establish a monopoly position which can be capitalized on by advertisers attempting to build up monopoly positions for products being advertised. Circulation is promptly translated into advertising revenue. In news, features, and direct advertising the large advertiser secures a consistent advantage. Large department stores occupy an important position as a source of revenue and in building up newspapers. By news men they have been called "the most sacred of sacred cows."[74] It has been said of the Chicago *Tribune*, "It is the ads for the $1.89 housedress that must meet the payrolls."[75] The consistent and heavy users of space are in a position to secure better advertising rates though their success will vary with numerous considerations including the policy of publishers. At one extreme, Munsey said "I will not quarrel with the sources of my revenue,"[76] and at the other, Scripps insisted on the necessity of constant squabbling with advertisers.[77] The dangers of dependence on a small number of large advertising patrons who favour a small number of papers compels constant attention to the problems of small advertisers. Department stores force down prices in certain advertised lines in order to secure a heavy traffic which will be directed to higher price goods.[78] The large newspaper securing newsprint under more advantageous circumstances and able to attract large advertisers provides a powerful stimulus to the production and sale of commodities with the most rapid turnover. Certain types of marketing organization such as the department store and certain types of urban communities, planned to give quickest access for the largest possible numbers to the marketing centre, are given direct encouragement. Urban architecture tends to be built around the store window.

The implications of the press in the twentieth century are suggested in the increasing importance of evening as contrasted with morning papers. Evening papers cater to individuals who have exhausted the possibilities of concentrated mental power and demand relaxation and entertainment rather than information and instruction.[79] Scripps directed the development of the evening paper to meet the neglected demands of industrial classes. The space devoted to sports news increased enormously.[80] Conservatism of monopoly in the morning papers favoured the advances in technique insisted upon in the evening papers by advertisers. Changes in type and format, adaptation of the colour press,[81] the introduction of faster presses effectively met the demands of advertisers in the evening and the Sunday papers and most conspicuously in the mail order catalogues.

In the United States the dominance of the newspaper was accompanied by a ruthless shattering of language, the invention of new idioms, and the sharpening of words.[82] In England the impact of this development was checked by the dominance of the book but the similarity of language favoured a rapid borrowing of technical developments from the United States.[83] Northcliffe took full advantage of American experience.[84] He drastically reduced the reports of debates and, followed by others, introduced, in the words of Blumenfeld, "one of the most significant changes"[85] in the history of British journalism. Following Pulitzer his rule was "Never attack an institution! Attack the fellow at the head of it!"[86] His success was evident in the eviction of heretical newspaper editors from British newspapers after 1900 and the decline of rational political journalism.[87]

The effects of the new journalism reflected in the demands of advertisers for circulation and excitement were particularly striking in foreign policy.

Complaints had been made from an early date that British correspondents on the Continent had been appointed for their linguistic ability, and their lack of journalistic sense had become evident particularly after the telegraph had superseded the post. On political grounds Beaconsfield had persuaded the *Standard* to make Abel its correspondent in Berlin in 1878.[88] The conventional freedom of the Foreign Office from discussion in Parliament and the tendency of the press to turn from domestic to foreign affairs, particularly after the Home Rule controversy, gave the *Times* a strong stabilizing position in which to attempt to build up friendly relations with Germany. Weakening of the *Times's* position as a result of encroachment of papers which had emerged following the abolition of taxes on knowledge and its loss of prestige over the Civil War[89] and the Irish question brought it to the point of bankruptcy in 1890 when Moberly Bell was appointed manager. As an organ of comment and criticism it had fallen behind in its news services. In the interests of economy Moberly Bell was compelled to rely on younger inexperienced men to build up its news staff.[90] In 1897 Greenwood described foreign correspondence as "the most difficult and least satisfactory service of the press in Britain."[91]

The activity of the new journalists became evident in the South African War when a small noisy group through control over the telegraph and the presses was able to foment friction. The editor of the *Cape Times* was a special correspondent of two London dailies and editors of leading Johannesburg journals were imported from England to support the policy of Rhodes. The difficulties of the *Times* gave Northcliffe an opportunity to exploit personalities in the growth of antagonism to Germany. Newspapers ceased to emphasize views important to France and became anti-German; it was in the words of Norman Angell "most profitable to increase that national danger which was already her greatest danger."[92] Lack of professional standards was responsible for the remarks of Sir Mountstuart Grant Duff: "The diplomatists and foreign ministers of Europe would get on perfectly well together and settle their own differences but for the new journalists intermeddling and stirring up international jealousy and spite."[93] The emphasis on personalities was evident in the claim that the resentment of Edward VII at criticism by the Kaiser of his participation in a week-end game of baccarat—the Tranby Croft episode—was the real beginning of the Anglo-French entente.[94] We can perhaps subscribe to Acton's remark that "nothing causes more envy and unfairness in men's view of history than the interest which is inspired by individual characters."[95] Control of the *Times* by Northcliffe[96] after 1908 facilitated a combination of prestige and circulation. The influence of American experience[97] is shown in the work of F. W. Wile, a correspondent of the

Chicago *Daily News*, and a temporary correspondent of the *Daily Mail*, who interviewed the British Secretary of War and gave the exclusive news to the *Daily Mail* for which he was rewarded by being placed in charge of the Berlin Bureau. Wile quotes the story that immediately prior to the outbreak of war a member of the Foreign Office in Berlin about to sign his passport "threw down the pen on the table, and said he absolutely refused to sign a passport for Wile because he hated him so and because he believed he had been largely instrumental in the bringing about of the war."[98] Whether we believe Wile or believe that he believed this, the remark is sinister. Norman Angell wrote that where the instructive modes of conduct were pretty evenly balanced, the power of an individual such as Northcliffe was decisive.[99]

The impact of the new journalism imported from the United States and adapted to Great Britain was even more conspicuous in the First World War. The role of Northcliffe in the reorganization of the Cabinet has been discussed at length and I shall be content with a comment by Mr. Winston Churchill after the Coalition. Northcliffe "wielded power without official responsibility, enjoyed secret knowledge without the general view, and distributed the fortunes of national leaders without being willing to bear their burdens."[100] But the new journalism brought a new type of politician. Frank Dilnot has described the friendly, informal way of Lloyd George with journalists, which "so far as Cabinet Ministers are concerned is something of a revolution."[101] H. W. Massingham commented on the change. "Unhappily for political society it breeds a dozen Lloyd Georges for one Courtney—our suicidal journalism understands only the first type and ignores or belittles the other."[102] The new type of politician had few scruples in using the press. Whether or not we accept the opinion of George Lansbury that Lloyd George would have ended the war at the time of the Russian Revolution if he thought he could have beaten Northcliffe and the *Daily Mail*,[103] it is clear that Lloyd George's control over the *Daily Chronicle*[104] became a measure of self-defense.

After the war Kennedy Jones admitted that the press had overemphasized its importance and power and that it had lost prestige and was distrusted by the public. Northcliffe[105] and the *Daily Mail* were followed by Beaverbrook and the *Daily Express*.[106] Interest in European politics was displaced by interest in Imperial preference. Large newspapers must have a foreign policy. The density and distribution of large centres of population in Great Britain, as contrasted with the United States, favoured the establishment of regional editions of the large London papers. The small number of papers with large circulation made a policy of control over newsprint mills and supplies of raw material more important than in the United States. In turn large aggregations of capital[107] were wholly concerned with circulation and advertising revenue. Along with insurance schemes and other devices to increase circulation the attempt to secure acceptance of British Empire preference occupied an important place. It would be dangerous to accept the claims of the newspapers concerned but Beverley Baxter wrote that Neville Chamberlain saw Beaverbrook and "a real peace was arranged based on a broad if not a precise acceptance of Beaverbrook's policy." "Beaverbrook won the battle of policy and lost the battle of personalities."[108] In any case free trade came to an end. As in the United States,[109] the monopoly position of a group of papers invited competition from others anxious to take advantage of improved technology and to serve neglected purchasers and readers. The phenomenal success of the *Daily Herald* in expanding circulation pointed to the limitations of established papers. Newspapers such as the *Morning Post*, unable to compete in the race for circulation, became "a parasite in the advertising business" and disappeared.[110] The claim that obsession with advertising and circulation in the period from 1919 to 1939 coincided with a lack of intelligent interest in public affairs,[111] and a lack of effective opposition to foreign policy at an early date, appears to have justification.[112] Foreign policy reflected the lumpy character of technological development and the monopolistic demands of newspapers.

The impact of Anglo-American journalism on Continental journalism was delayed as a result of differences in language and the stronger position of the book.[113] French newspapers were less subject to the overwhelming influence of the advertiser[114] and more subject to the influence of direct subsidy shown in the effects of Russian[115] and Italian bribery. Political and other writing was more important than news. Journalists played a more important role in political life. A common proverb prevailed that journalism led to everything provided one got out of it. The German press retained the characteristics it had developed under the tutelage of Bismarck. A student of the press has referred to "the poisoning of public opinion by the newspaper press in all the great countries."[116] Kent Cooper, manager of the Associated Press, stated that "the mighty propaganda carried on through these channels in the last hundred years has been one of the causes of war that has not yet been uncovered."[117] But European civilization was still dominated by the book, and war between Germany and Anglo-Saxon countries could be described as a clash between the book and the newspaper. The weakness of newspapers in Germany probably accentuated the power of propagandist organizations. Germany was unable to appreciate the power of the newspaper in Anglo-Saxon countries, and collapse was in fact a result of increasing difficulties of understanding incidental to differences in development of the newspaper in the two regions. By the newspaper, democracy had completely expelled the book from the normal life of the people.[118] One is perhaps in danger of accepting at face value the description of the work of propagandists during the war from 1914 to 1918 but undoubtedly much can be attributed to the work of Northcliffe, "a master of mass suggestion."[119] Advertising technique had not developed in Germany to the point that propaganda could be effectively used or effectively resisted.

The Treaty of Versailles reflected the influence of the printing press[120] by its emphasis on self-determination as a governing principle but it neglected the importance of the spoken word upon which the radio capitalized.[121] Limitations of the press in Germany facilitated the rapid development of the radio in contrast with restrictions imposed on it in England through the influence of the press with its interest in advertising. Regions dominated by the German language were powerfully influenced by Hitler in his efforts to extend the German Reich. In regions dominated by the English language during the war resources were mobilized more effectively by the speeches of Churchill and Roosevelt. Dissimilarities in language were an important factor in the development of underground movements in occupied countries. The Russian language proved an effective barrier to German propaganda. With the radio and events changing dramatically from moment to moment, people of the same language began to rely on it for information and with the possibilities of control of technical equipment it became a powerful instrument for propaganda.[122] If the First World War might be regarded as a clash between the newspaper and the book, the Second World War was a clash between the newspaper and the radio.

Technological advance in communication implies a narrowing of the range from which material is distributed and a widening of the range of reception, so that large numbers receive, but are unable to make any direct response. Those on the receiving end of material from a mechanized central system are precluded from participation in healthy, vigorous, and vital discussion. Instability of public opinion which follows the introduction of new inventions in communication designed to reach large numbers of people is exploited by those in control of the inventions. Hearst has advised that the important thing for a newspaper to do in making circulation is to get excited when the public is excited.[123] The lumpy character of new developments in communication and the appeal to lower levels of intelligence by unstable people accentuates instability if not insanity. It is only necessary to recall the effects of the radio as shown in the Munich crisis and especially in Orson Welles's "Invasion from Mars." In the attempt to reach larger numbers, a new type of skill on the part of the writer becomes necessary; in contrast with the monotony of hard labour which leaves more time to think, the fatigue and monotony of the new type of brain work leave exhaustion.

As a result of the emphasis on simplicity on the part of the writer, the problems of government become more complex. Large numbers of professional organizations and dispensers of propaganda learned their trade during the First World War.[124] "The epidemic itch for manipulating the public has infected the population in a rash of press agents, publicity experts, advertisers, and propagandists."[125] As a result the role of the civil service becomes more important and the position of bureaucracies is strengthened. Dicey's remark that "laws . . . are . . . among the most potent of the many causes which create public opinion"[126] has gained in weight. Even Kennedy Jones conceded the power of government as a source of news and deplored the development of an agency system in Downing Street in which the agent was chosen for his ability to gauge correctly the effects of suppression of truth. Governments with their control over news can exercise what is called leadership. To give the appearance of maintaining control over policy, newspapers are compelled to keep very close to government leaders and are precluded from criticism. To appear consistent the newspaper is compelled to adopt a broad policy which will permit change without the appearance of change and the possibility of unostentatiously taking curves.

We have suggested that the enormous technological advance at various stages in the production of the newspaper has been supported by the increasing power of advertising reflected in changes in the character of news, features, and editorial opinions. The new journalism emphasized a vast range of interests at the expense of politics and with the rise of public relations agencies, lost the power to expose abuses, particularly abuses from which it gains. As a result of its interrelation with news, features, and editorial opinion, advertising became monopolistic in relation to a monopolistic press and imposed its influence on political, social, and economic life. The consequent maladjustments were evident in the boom of the twenties and the depression and were to an important extent a result of expansion of the press and of a new instrument of communication—the radio. Public opinion became less stable and instability became a prime weakness serving as a forced draft in the expansion of the twenties and exposed to collapse in the depression.

The highly sensitive system created in advertising through the press faced disaster and compelled large-scale state intervention. We can appreciate Thorold Rogers's remarks that "variations of high and low prices, which a century ago would have excited little attention and caused little alarm, in our day when production and trade are so sensitive and so complicated rouse the gravest apprehensions and exercise the attention of the most laborious and acute investigators into economical phenomena and economical agencies."[127] Government intervention implied the growth of nationalism. Advertising through the press contributed directly and indirectly to the growth of tariffs except as we have seen in such raw materials as newsprint used by advertisers, and the inadequacies of tariffs in the depression were met by manipulation of exchange rates and concern with monetary control with results suggested by Professor Robbins in his analysis of the great depression. Maine's remark that "energy poured into the study of economics and especially of money leads to the neglect of politics and government"[128] has been amply underlined. Economic writing and discussion have become increasingly nationalistic, concerned with specific legislation and specific economic and social trends. To paraphrase Mr. Dooley, when the whole world goes crazy the social scientist must go crazy too.

The weakness of the social sciences has been shown in the increasing concern with national statistics. Von Beckerath has stated that "truth in the social sciences is, in the last analysis, often nothing but the consistency and conformity of facts, endeavours, and ideas with the basic tenets of the respective society.[129] For example, concentration on study of the business cycle has possibly influenced the course of business cycles in the sense that nature copies art. The creation of a belief in the business cycle may lead to action on the part of business men designed to respond to the business cycle and possibly intensify, possibly check, the violence of booms and depressions. Nations which have not had the advantage of special studies cycles will probably have different types of business cycles. Professor Ohlin has complained of the insularity of Anglo-American economic thought in the neglect of Wicksell's work for a period of two decades, but he forgets the bias of economics which makes the best economists come from powerful countries. The influence of economists in special studies is reinforced by their active role in business and government. Business men and governments are apt to insist on hearing and reading what they want to hear and read and consequently are provided with what they want to hear and read.

Nationalism weakens an interest in universal laws. Similarity in patterns of development in administrative techniques within nations are misinterpreted as laws. The results are evident in the character of writing in the social sciences in this century. A letter from Hume to Adam Smith contains the sentence: "Nothing, indeed, can be a stronger presumption of falsehood than the approbation of the multitude,"[130] and this opinion persisted in such remarks in the nineteenth century as those of G. C. Lewis: "The concurrence of the crowd is a proof of the worst side,"[131] or of George Sand: "There is nothing so undemocratic as the mass of the people." The necessity of a sustained interest in the social sciences was reflected in a comment of Christopher North that "political economy is not a subject for a magazine"[132] and of

Alfred Marshall, "you cannot afford to tell the truth for half a crown."[133] The contrast in the twentieth century became evident in the belief of Keynes that it could be sold for two half-crowns. The tradition of Marshall's monopoly of teaching and his reluctance to publish was continued by Pigou and led to a disequilibrium of economic thought and the outpouring of Keynesian literature. The decline in influence of the politician in the press was shown in the political failure of the Versailles Treaty and the success of Keynes's *Economic Consequences of the Peace* (1920).

The limitations of these new attempts to offset the influence of nationalism have been suggested in Mantoux's *Economic Consequences of Mr. Keynes* (1946) in which it is argued that the point of view of Anglo-Saxon law and its concern with trade and commerce ignored the point of view of continental and Roman law and its concern with political and military considerations. The weakness of the social sciences as a result of the obsession with national statistics became evident in this neglect of differences which had become more acute with the later development of the press on the Continent. The impact of science on industrialism in England reduced the possibility of speculative interest in law peculiar to the Continent. The rigid concept of property in Roman law failed to provide a basis for the growth of trade and of political economy, in contrast with the elastic concept of ownership in common law. In common law countries, political economy emerged to meet the problems of law in industry and commerce. Under the influence of law, Adam Smith extended his speculation to political economy but his tradition was not sufficiently strong to resist the influences of nationalism. The interrelations of law and political economy have been noted by Sir Henry Maine.

It is certain that the science of Political Economy, the only Department of moral inquiry which has made any considerable progress in our day, would fail to correspond with the facts of life if it were not true that Imperative Law had abandoned the largest part of the field which it once occupied, and had left men to settle rules of conduct for themselves with a liberty never allowed to them till recently. The bias of most persons trained in political economy is to consider the general truth on which their science reposes as entitled to become universal, and, when they apply it as an art, their efforts are ordinarily directed to enlarging the province of Contract and to curtailing that of Imperative Law, except so far as law is necessary to enforce the performance of Contracts. The impulse given by thinkers who are under the influence of these ideas is beginning to be very strongly felt in the Western World. Legislation has nearly confessed its inability to keep pace with the activity of man in discovery, in invention, and in the manipulation of accumulated wealth; and the law even of the least advanced communities tends more and more to become a mere surface-stratum, having under it an ever-changing assemblage of contractual rules with which it rarely interferes except to compel compliance with a few fundamental principles, or unless it be called in to punish the violation of good faith.[134]

Political economy as an extension of law was accompanied by an enormous advance in order and in industry and commerce in the nineteenth century and had the advantages and limitations of law. "The less of science claimed for law, the greater the element of justice dispensed in its administration. The more the law seeks formal objectivity the less justice it may be feared will be strained out."[135] "Social things do not lend themselves to precision and whatever principles we get that are precise do not lend themselves to social things."[136] "If the social sciences as sciences formulate themselves so analytically and autonomously as to rise above custom and gossip they cease to be social. If they remain social they will be so involved in the medley of life as no longer to be scientific in terms of the indicated precision."[137] Bagehot wrote that "the practical value of the science of political economy . . . lies in its middle principles."[138] "No social science of any department is decisive in the sense of being in a position to dictate to us the necessary or the best lines of conduct."[139]

The effects of nationalism shown in a concern for aggregates, estimates, and averages has accentuated a narrowing interest in mathematical abstractions and a neglect of the limitations of precision. Maurice Dobb has complained of the obsession with algebraic symbols and the neglect of the value of labour. European writers such as Pareto who did important work in mathematics round out their studies by treatises in sociology. Marshall has commended Bentham for stressing the importance of measurement. "When you have found a means of measurement you have a ground for controversy and so it is a means of progress,"[140] but there have been signs that it has become a basis of immodest finality. We are reminded of Pliny's comment that magic embraced religion, mathematics, and medicine, the three arts that most rule the human mind. We would do well to remember the statement of Gibbon: "As soon as I understood the principles I relinquished forever the pursuit of mathematics, nor can I lament that I desisted before my mind was hardened by the habit of rigid demonstration, so destructive to the finer feelings of moral evidence." Hallam wrote that "one danger of this rather favourite application of mathematical principles to moral probabilities, as indeed it is of statistical tables (a remark of far wider extent) is, that, by considering mankind merely as units, it practically habituates the mind to a moral and social levelling, as inconsistent with a just estimate of man as it is characteristic of the present age."[141]

The neglect of law and of qualitative considerations has accentuated an interest in the price system with the result that the impact of advertising through the press on the social sciences has been overwhelming. The powerful influence of the logic of mathematics and the natural sciences has contributed to the rigidity of law and increased the complexity of legislation and the power of bureaucracy and authoritarianism. The cumulative effects have been evident in the crowding of weaker students from mathematics and the natural sciences into the social sciences where their slight knowledge of mathematics gives them an advantage. It is perhaps the cumulative bias of mathematics which has led Professor T. V. Smith to remark that whereas growth is continuous, social science seems almost to shift base from time to time rather than to grow from more to more upon the same trunk.[142]

Perhaps the most serious results of these tendencies are shown in the lack of interest among social scientists in other civilizations than those of the west, in the neglect of philosophical problems, and in the obsession with scholastic problems of reconciling dynamic and static theories. The Chinese concept of time, for example, as plural and characterized by a succession of times, which reflects their social organization with its interest in hierarchy and relative stability, as well as their concept of space, has been adapted through collective collaboration and experience to social life. The Western concept of time with its linear character, reinforced by the use of the decimal system, has in contrast a capacity for infinite extension to the past and the future and a limited capacity for adaptation.[143] Our study of the press has suggested that insistence on time as a uniform and quantitative continuum has obscured qualitative differences and its disparate and discontinuous character. Advertisers build up monopolies of time to an important extent through the use of news. They are able to take full advantage of technological advances in communication and to place information before large numbers at the earliest possible moment. Marked changes in the speed of communication have far-reaching effects on monopolies over time because of their impact on the most sensitive elements of the economic system. It is suggested that it is difficult to over-estimate the significance of technological change in communication or the position of monopolies built up by those who systematically take advantage of it. The disequilibrium created by the character of technological change in communication strikes at the heart of the economic system and has profound implications for the study of business disturbance. One might dilate on the implications of varying rates of development for the problem of international understanding and on the effects of rapid extensions of communication facilities on instability and the savagery of war, as Liddell Hart has done in comparing the savagery in the period following the invention of printing and that of the present century. Freedom of the press as guaranteed by the Bill of Rights in the United States has become the great bulwark of monopolies of time. The results of the American Revolution hang heavily over the world's destiny. It should be clear that improvements in communication tend to divide mankind, and perhaps I may be excused for ending this lecture without encroaching on copyright by using the words of a title of one of Aldous Huxley's novels, *Time Must Have a Stop*.

Notes

1. J. E. Thorold Rogers, *The Economic Interpretation of History* (New York, 1888), p. 307.
2. Ibid., p. 339.
3. It replaced the speaking tube at an early date.
4. S. G. Blyth, *Making of a Newspaper Man* (Philadelphia, 1912), p. 184.
5. The opposition to control of newspapers by mills does not emerge in the case of the control of mills by newspapers. Newspaper chains were in part a result of a tendency toward trusts in the manufacture of newsprint.
6. M. Koenigsberg, *King News* (New York, 1941), p. 397.
7. George Seldes, *Freedom of the Press* (New York, 1935).
8. H. M. Hughes, *News and the Human Interest Story* (Chicago, 1940).
9. J. W. Linn, *James Keeley, Newspaperman* (Indianapolis, 1937).
10. Charles Mertz, *The Great American Bandwagon* (New York, 1928), p. 71.
11. R. D. Blumenfeld, *R.D.B.'s Diary, 1887–1914* (London, 1930), p. 137.
12. Herbert Stone, the son of Melville Stone, a newspaperman, attempted to establish publishing on an effective footing. He republished much material appearing in newspapers, for example, the writings of George Ade. Sidney Kramer, *A History of Stone and Kimball* (Chicago, 1940).
13. Elmer Ellis, *Mr. Dooley's America: A Life of Finley Peter Dunne* (New York, 1951), p. 32. Storey as a printer was especially concerned with display type for heads and with appearance. F. B. Wilkie, *Personal Reminiscences of Thirty-five Years of Journalism* (Chicago, 1891), p. 119.
14. Seldes, *Freedom of the Press*. "As for Mr. Hearst's press . . . aside from the sporting and theatrical pages, I think they make the whole thing up" (Ilka Chase, *Past Imperfect* [New York, 1942], p. 5).
15. Seldes, *Freedom of the Press*.
16. Koenigsberg, *King News*, p. 207.
17. Cited by Hughes, *News and the Human Interest Story*.
18. F. A. Mackenzie, *Beaverbrook* (London, 1931), pp. 177–78.
19. R. D. Blumenfeld, *R.D.B.'s Procession* (New York, 1933), p. 137. See Alphonse Courlander, *Mightier Than the Sword* (London, 1913).

He remembered a story that Willoughby told of a Salvation Army meeting in the Albert Hall, when General Booth had walked up and down the platform speaking of the glories of salvation, and, suddenly, he pointed a finger at the table below. "Are you saved?" he asked, with his finger shaking at a man who was looking up at him. "Me?" said the man looking about him confusedly, and then, with a touch of indignation at being suddenly dragged into the game, "Me? I'm a reporter." (ibid., p. 108)

20. Arnold Bennett, *The Truth about an Author* (n.p., n.d.), p. 86.
21. H. O. Mahin, *The Development and Significance of the Newspaper Headline* (Ann Arbor, 1924), p. 148.
22. Arnold Bennett, *Journalism for Women* (London, 1898), p. 4.
23. Ishbel Ross, *Ladies of the Press* (New York, 1936), p. 108.
24. Cited in Denys Thompson, *Voice of Civilization* (London, 1943), p. 111.
25. T. H. S. Escott, *Masters of English Journalism* (London, 1911).
26. Harper Leach and John C. Carroll, *What's the News?* (Chicago, 1926).
27. F. W. Wile, *News Is Where You Find It* (Indianapolis, 1939), p. 173.
28. Cited in Edward Cook, *Delane of the Times* (London, 1915), p. 197.
29. Cited in H. W. Boynton, *Journalism and Literature* (Boston, 1909), p. 13.
30. Hamilton Holt, *Commercialism and Journalism* (Boston, 1909), p. 4.
31. W. F. Johnson, *George Harvey* (Boston, 1929), p. 98.
32. George Britt, *Forty Years—Forty Millions: The Career of Frank A. Munsey* (New York, 1935), p. 185.
33. William Allen White, *The Autobiography of William Allen White* (New York, 1946), p. 629.
34. Philip Gibbs, *Adventures in Journalism* (New York, 1923).
35. Louis Cazamian, *Criticism in the Making* (New York, 1929), p. 78.
36. D. C. Seitz, *Joseph Pulitzer, His Life and Letters* (New York, 1924), p. 406.
37. H. I. Brock, *Meddlers: Uplifting Moral Uplifters* (New York, 1930), p. 276.
38. Wordsworth commented on "Illustrated Books and Newspapers" in 1846:

> Discourse was deemed Man's noblest attribute,
> And written words the glory of his hand;
> Then followed Printing with enlarged command
> For thought—dominion vast and absolute
> For spreading truth, and making love expand.
> Now prose and verse sunk into disrepute
> Must lackey a dumb Art that best can suit
> The taste of this once-intellectual Land.
> A backward movement surely have we here,
> From manhood—back to childhood; for the age—
> Back towards caverned life's first rude career.
> Avaunt this vile abuse of pictured page!
> Must eyes be all in all, the tongue and ear
> Nothing? Heaven keep us from a lower stage!

39. S. M. Bessie, *Jazz Journalism* (New York, 1938), p. 236.
40. Burton Rascoe, *Before I Forget* (New York, 1937), p. 276. In England newspaper readers wanted blood, sports, religion, birth control, vivisection, household budgeting, spiritualism, and should married women work (Reginald Pound, *Their Moods*

and Mine [London, 1939], p. 146). Spiritualism attracted wide interest: "If I am remembered at all a hundred years hence, it will be as Julia's amanuensis" (E. K. Harper, *Stead, the Man: Personal Reminiscences* [London, 1914], p. 1). "It's piety that pays, especially when it's printed" (H. Rider Haggard, *Mr. Meeson's Will* [London, 1888]).

41. Emile Gauvreau, *My Last Million Readers* (New York, 1941), p. 221.
42. Ibid., p. 177.
43. Emile Gauvreau, *Hot News* (New York, 1931), p. 290.
44. Ibid., p. 31.
45. Ibid., p. 98.
46. Wilbur Forest, *Behind the Front Page* (New York, 1935), p. 310.
47. K. Stewart, *News Is What We Make It* (Boston, 1943), p. 106.
48. W. S. Maulsby, *Getting the News* (New York, 1925), p. 27.
49. M. M. Willey, *The Country Newspaper* (Chapel Hill, 1926), p. 93.
50. The Ideal Home Exhibition of 1908 emphasized the *Daily Mail* as a home paper (Tom Clarke, *Northcliffe in History* [London, n.d.], p. 99). Serial stories were designed to attract women and children's stories strengthened circulation (ibid., p. 116).
51. S. A. Moseley, *The Truth about a Journalist* (London, 1935), p. 304.
52. Mark Sullivan, *Education of an American* (New York, 1938).
53. C. L. Edson, *The Gentle Art of Columning* (New York, 1920), p. 120.
54. Hearst was said to have supported Bryan partly because of the silver properties in his father's estate. He seemed "to think that democracy is an end in itself and that the end justifies the means—his journalism" (Lincoln Steffens) (Lloyd Morris, *Postscript to Yesterday* [New York, 1947], pp. 236–37). The political changes of Chicago papers and losses in circulation opened the way for Hearst. Kohlsaat bought the Chicago *Post and Mail* from J. R. Walsh and changed the policy from Democratic to Republican; Walsh in turn purchased the *Interocean* and changed its name to the *Chronicle* and its policy from Republican to Democratic.
55. They opposed the position of the Associated with its interest in monopoly and free enterprise and emphasized municipal reform—for example in Cleveland. Ability to buy paper for as low as $1.65 enabled them to start one-cent papers. With a detailed knowledge of the press they looked over cities of 250,000 or more and introduced machines capable of producing 36,000 48-page papers per hour. They were compelled, and in this secured the co-operation of department stores, to take active measures to introduce lower coinage denominations. They exploited the weakness of political city machines and their party organs, hence gaining publicity as a result of their fights and a reputation for independence. After 1903 they introduced a vigorous campaign against proprietary medicines and opened the way for foreign advertising. They followed a consistent policy of enlisting the financial interest of young men by making them stockholders and starting new papers, as in Toledo and Columbus, to provide employments. See M. A. McRae, *Forty Years in Newspaperdom* (New York, 1924).
56. Delos Wilcox, cited in S. M. Kingsbury, Hornell Hunt, and associates, *Newspapers and the News* (New York, 1937), p. 199.
57. K. A. Bickel, *New Empires* (Philadelphia, 1930), p. 39.

58. Kingsbury et al., *Newspapers and the News*, p. 158. "We don't direct the ordinary man's opinion, we reflect it." "If it's flattery to show the man-in-the-street his own importance, then we are flatterers" (Northcliffe).

59. W. J. Bryan and M. B. Bryan, *The Memoirs of William Jennings Bryan* (n.p., n.d.), p. 299.

60. *The Changing Years: Reminiscences of Norman Hapgood* (New York, 1930), p. 177.

> Cautiously replied the beaver,
> With discretion made he answer
> Give me time to ask the others
> Let me ask the other beavers.
>   Henry Wadsworth Longfellow, *Hiawatha: A Poem*

61. C. R. Corbin, *Why News Is News* (New York, 1928). "News is what a chap who doesn't care much about anything wants to read" (Evelyn Waugh, *Scoop: A Novel about Journalists* [London, 1948], p. 69).

62. Leech and Carroll, *What's the News?*

63. Press agents recruited from the ranks of newspaper reporters have a wide range of information, are keen observers and shrewd judges of men and know what is interesting. Their advice is sought by editors, and others looking for special knowledge. J. E. Hedges, *Common Sense in Politics* (New York, 1918). "It is carefully noted that up to the age of ten they [the politicians] had never tasted whiskey and from then to the age of fifteen they had never struck their parents. . . . The public is quietly informed that in early years he went without food to buy books." "The first impressions gained by the public are apt to be lasting" (ibid., pp. 126–27).

64. Will Irwin, *The Making of a Reporter* (New York, 1942), p. 165.

65. H. L. Stoddart, *As I Knew Them: Presidents and Politicians from Grant to Coolidge* (New York, 1927), p. 553.

66. Frank Kent, *The Great Game of Politics* (New York, 1940), p. 93.

67. George Seldes, *Lords of the Press* (New York, 1938), p. 238.

68. James A. Farley, *Behind the Ballots: The Personal History of a Politician* (New York, 1938), p. 319. See also *Jim Farley's Story* (New York, 1948) for an illuminating account of the decline of the Postmaster-General as distributor of patronage.

69. Cited in Raymond Moley, *After Seven Years* (New York, 1939).

70. Hedges, *Common Sense in Politics*, p. 156. "The grand result of these and a myriad other circumstances is a fevered condition of mind, a passion of haste, desperate competition and accomplishment, adverse to deliberation, preventive of calmness of judgment, and conducive to loss of perspective" (ibid., p. 51).

71. Whitelaw Reid, *American and English Studies* (New York, 1913), pp. 220–21.

72. N. B. Mavity, *The Modern Newspaper* (New York, 1930), p. 11.

73. H. D. Lasswell, *Propaganda Technique in the World War* (London, 1927).

74. Seldes, *Lords of the Press*, p. 156.

75. H. L. Ickes, *Freedom of the Press Today* (New York, 1951), p. 87. In 1928–1929 London papers became more definitely "drapers' circulars" (St. John Ervine, *The Future of the Press* [n.p., n.d.], p. 7).

The Press, a Neglected Factor in Economic History 99

76. Britt, *Forty Years—Forty Millions*, p. 197.
77. N. D. Cochran, *E. W. Scripps* (New York, 1933).
78. Jason Rogers, *Newspaper Building* (New York, 1918).
79. W. G. Bleyer, *Main Currents in the History of American Journalism* (Boston, 1927), p. 390.
80.
> The illustrated papers that are read, apart from serious news, are a revelation of the vacuity of the public mind, as the advertisements are a testimony to its imbecility. The absence of any thoughts or information that can enlarge the mind, or give it fresh insight or understanding, and the fatuity of the illustrations, show the helpless little round of common ideas of the well-to-do classes: while the dishing up of legal filth for the lower classes, and the morbid love of trivial accidents and catastrophes, shows terribly the mere animalism which fills their horizon. The one subject on which most print is spent is that which is absolutely futile, sport and games. Whether one group of men, selected by mere accident, is a minute trifle more active than another accidental group, is a matter of such utter insignificance that it would seem impossible to suppose that anyone would turn the head to see the result decided. Yet such questions absorb most of the interests and spare thoughts and reading of a great part—perhaps the greater part—of the population, just as the races of the circus swamped all other interests of the decadent Roman. The results which they crave for cannot possibly mean anything to the present or to the future, as the selection is merely due to accidental causes. Even a lower depth is the relative excellence of two horses which are completely unknown to the persons who speculate on them. The utter waste of thought and print in such interests is a form of insanity which is worse than a drug habit, as it implies a hopeless atrophy of the mind to interests which would help it or develop it. (W. M. Flinders Petrie, *Janus in Modern Life* [New York, 1907] p. 18).

See also George Blake, *The Press and the Public* (London, 1930).
81. Edward W. Bok, *A Man from Maine* (New York, 1923), p. 172.
82. W. M. Fullerton, *Problems of Power* (New York, 1913), p. 21.
83. Morley wrote: "For a newspaper must live, and to live it must please, and its conductors suppose, perhaps not altogether rightly, that it can only please by being very cheerful towards prejudices, very chilly to general theories, disdainful to the men of principle" (*On Compromise* [London, 1923], p. 22).

"It is, however, only too easy to understand how a journal, existing for a day, should limit its view to the possibilities of the day, and how, being most closely affected by the particular, it should coldly turn its back upon all that is general. And it is easy, too, to understand the reaction of this intellectual timorousness upon the minds of ordinary readers, who have too little natural force and too little cultivation to be able to resist the narrowing effect of the daily iteration of shortsighted commonplaces" (ibid., p. 22).

"The education of chiefs by followers, and of followers by chiefs, into the speedy abandonment of the traditions of centuries or the principles of a lifetime may conduce to the rapid and easy working of the machine. It marks a triumph of the political spirit which the author of *The Prince*, Machiavelli himself, might have admired" (ibid., p. 78).

84. Alfred Harmsworth and others were raised to the peerage partly as a result of the need to advertise the popular press. The *Manchester Guardian* was able to advertise as "the press without a peer." Harmsworth had a large printing plant built up for

the production of *Answers* and with American salesmanship methods he weakened the position of Cassells. He had enormous sales of Harmsworth's encyclopaedia and developed the children's encyclopaedia which became the *Book of Knowledge*. See J. A. Hammerton, *Books and Myself* (London, 1944).

85. Ralph David Blumenfeld, *The Press in My Time* (London, 1933), p. 113.

86. Valentine Williams, *The World of Action* (Cambridge, 1938), p. 138.

87. G. P. Gooch, *Life of Lord Courtney* (London, 1920), p. 404. The *News* under Quaker domination probably suffered form lack of flexibility incidental to insistence on principles. See A. G. Gardiner, *Life of George Cadbury* (London, n.d.). Northcliffe "killed the penny dreadful by the simple process of producing a ha'penny dreadfuller" (A. A. Milne).

88. Charles Lowe, *The Tale of a Times Correspondent* (London, n.d.), p. 98.

89. Contact with American journalism was evident in the planting of news unfavourable to the north to be printed in American papers and used in favour of the south (R. D. Harper, *Lincoln and the Press* [New York, 1951], pp. 98–99). The *Times* report on the Gettysburg address stated that "the ceremony was rendered ludicrous by some of the sallies of that poor President Lincoln."

90. E. H. C. Moberly Bell, *Life and Letters of C. F. Moberly Bell* (London, 1927), p. 309.

91. Cited in L. M. Salmon, *The Newspaper and the Historian* (New York, 1923), p. 188.

92. Norman Angell, *The Press and the Organization of Society* (London, 1922), p. 39.

93. Escott, *Masters of English Journalism*, p. 338.

94. Irwin, *The Making of a Reporter*, p. 270.

95. F. E. Lally, *As Lord Acton Says* (New York, 1942), p. 211.

96. The new journalism compelled the *Times* to emphasize printing and led to its clash with the organized booksellers under the net book agreement.

97. Frank Taylor, *The Newspaper Press as a Power Both in the Expression and Formation of Public Opinion* (Oxford, 1898); A Believer, *Wanted Press Reform, a British Press for the British People, American for the Americans* (1906). A discussion of the claims of editors and publishers of newspapers of their powers would include a reference to Greeley and the emancipation of the slaves and to Pulitzer in the campaign to compel the government to sell bonds directly to the public. Biographies and autobiographies suggest a tendency toward exaggeration. John Evelyn Wrench, *Uphill* (London, 1934), p. 225, states that Northcliffe showed signs of becoming obsessed with power in 1909 by allowing his name to appear in his papers. According to McNair Watson he believed he had changed favourable opinion in England from Germany to France. He complained of the dominance of business in the newspaper field but neglected his own interest in the newspaper, the most dominant of all. Northcliffe had a "lack of character" (p. 199) but endless bounce. "For years news-editor of the *Daily Mail* before becoming Editor, W. G. Fish had a regular morning telephone call from the Chief. Sometimes contact was established by the mischievously worded, but apparently artless question, 'Anything fresh, Fish?'" (Bernard Falk, *He Laughed in*

Fleet Street ([London, 1937], p. 221). In the words of Leicester Harmsworth, there was "a touch of brutality about journalism."

Newspapers seem to be built up by instability and in turn to produce instability and the beginnings of movements as in the case of Sir Evelyn Wrench and the English-speaking Union, Norman Angell and the Great Illusion, and W. T. Stead. Dominance of newspapers by individuals assumes a system of spies and informers to keep them informed of details. Northcliffe probably followed Pulitzer (Alleyne Ireland, *Joseph Pulitzer: Reminiscences of a Secretary* [New York, 1914]). "Next to the Kaiser, Lord Northcliffe has done more than any other living man to bring about the war" (*The Daily Mail and the Liberal Press: A Reply to Scaremongerings and an Open Letter to Lord Northcliffe* [1914]). He was a man ever ready to set the world in a blaze to make a newspaper placard (p. 12). See also *After All: The Autobiography of Norman Angell* (London, 1951); W. L. George, *Caliban* (New York, 1920); Louise Owen, *The Real Lord Northcliffe: Some Personal Recollections of a Private Secretary, 1901–1922* (London, 1922); F. B. Wilkie, *Personal Reminiscences of Thirty-five Years of Journalism* (Chicago, 1891); [H. Ainsley] *Danger! or The Press and Its Would-be Napoleons, Their Hypocrisy and Future with Special Reference to Horatio Bottomley: A Warning and an Appeal to All;* by a Worker (London, 1922). The press had a vested interest in a strong opposition and could time its hardest blows when authority was in confusion, thus creating discord in the government, and showing its powers most effectively. "Every extension of the franchise renders more powerful the newspaper and less powerful the politician" (Northcliffe). It may be questioned whether women suffrage supported this view. For a valuable discussion see G. V. Ferguson, *John W. Dafoe* (Toronto, 1948).

98. Wile, *News Is Where You Find It*, p. 307.

99. Angell, *The Press and the Organization of Society*, p. 33.

100. Winston S. Churchill, *The World Crisis 1916–1918*, I (London, 1927), 245.

101. Frank Dilnot, *The Adventures of a Newspaper Man* (London, 1913), p. 251. Lloyd George gave Northcliffe a scoop on the Roads Bill in 1909 (Tom Clarke, *Northcliffe in History* [London, 1950], p. 88). On the other hand Northcliffe checked political bribery by raising wages.

102. H. J. Massingham, *A Selection from the Writings of H. H. Massingham* (London, 1925), p. 75. "The discovery of advertising was bound ultimately to revolutionize the press by changing the balance of power from the directly political to the directly commercial." See Wareham Smith, *Spilt Ink* (London, 1932). Northcliffe destroyed the hegemony of the editorial department with its pretensions of interest in public welfare. See C. E. Montague, *A Hind Let Loose* (London, 1918) on the futility of editorial writing. When the public felt that the newspaper was no longer interested in them it turned to the political party. The emphasis of newspapers on the short run and on consumption of goods for advertising possibly accentuated a trend towards nationalization and was made more important as a device for guaranteeing full employment to producers' industries. The importance of sport in the increase of circulation was evident in Hulton's success in the north with a sheet of racing news (A. M. Thompson, *Here I Lie:*

The Memorial of an Old Journalist [London, 1937], p. 41). Blatchford and Thompson as writers on sport papers and on the *Clarion* developed an informality of style. Their support of socialism meant an evasion of the abstractions of Marx. English socialism became a product of education, art, the vernacular, and the new journalism. The class war centred around propriety and impropriety rather than property and poverty. Labour objected to Blatchford and Thompson writing for Northcliffe but to the latter Blatchford was "more like an old Tory squire than a modern journalist." Blatchford's letters on Germany and the election campaign of 1910 which described the Germans as consumers of black bread as a result of their protection system had serious effects on opinion (E. C. Bentley, *Those Days* [London, 1940], pp. 209–11). Blatchford adapted the language to those who had been educated under the act of 1870. Though John Burns described him as "a yellow press scribe lying like a gas meter" he could say "I can't speak but I can write." He brought the lower classes to the surface in political change.

103. *My Life* (London, 1931), p. 212.

104. It was stated that Sir Andrew Weir (Lord Inverforth) bought the *Daily Chronicle* for £1,600,000 and next day became Minister of Munitions (*They Told Barron*, ed. and arranged by Arthur Pound and S. T. Moore [New York, 1930], p. 200).

105. It is interesting to speculate on developments if Northcliffe had lived. "I suppose the Lord created Beaverbrook to make Rothermere tolerable" (Sir John Pollock, *Time's Chariot* [London, 1950], p. 72). Following Northcliffe the danger of great names led to the formation of a trust. It has been held that this gave the editor far too much irresponsible power.

106. Sidney Dark, *The Life of Sir Arthur Pearson* (London, n.d.).

107. See *Printers, Press and Profits* prepared by W. Fox for the Labour Research Department, Nov. 1932; also Lord Camrose, *London Newspapers: Their Owners and Controllers* (London, 1939).

108. Beverley Baxter, *Strange Street* (New York, 1935), pp. 252–53.

109. On the other hand the absolute indefeasible right of free speech was not recognized in England and the power of unions was such as to raise questions of policy. See H. E. B. Ludlam, *Industrial Democracy and the Printing Industry* (Coventry, 1924). Labour was reluctant to support nationalization of the press as it would involve bureaucratic dictatorship but it favoured such control over the press as would check attacks on trade unions.

110. Wilfrid Hindle, *The Morning Post, 1772–1937* (London, 1937), p. 240.

111. It was claimed that despatches of foreign correspondents were mutilated because they were contrary to received opinion and that war might have been avoided if the British correspondents of *The Times* and the *Daily Telegraph* had been taken seriously. See Pollock, *Time's Chariot*, pp. 239–40.

112. Denys Thompson, *Voice of Civilization* (London, 1943), pp. 126–27, 194.

113. H. M. Chadwick, *The Nationalities of Europe and the Growth of National Ideologies* (Cambridge, 1945); see also Vincent Sheean, *Personal History* (New York, 1940).

114. See Raymond Gillet, *Quelques idées et moyens de propagande pour la diffusion d'un journal* (Lyon, 1932).

115. See René Marchand, *Un Livre Noir* (Paris, n.d.).
116. J. F. Scott, "The Press and Foreign Policy," *Journal of Modern History*, Dec. 1931, p. 629.
117. Kent Cooper, *Barriers Down* (New York, 1942), p. 8.
118. Oswald Spengler, *The Decline of the West* (London, 1922), pp. 460 ff.
119. Sir Campbell Stuart, *Secrets of Crewe House: The Story of a Famous Campaign* (London, 1920), p. 130.
120. Language became more important as a national factor after the first world war (Michells). "Remember what the short-view press did to make the Treaty of Versailles inevitable" (Ivor Brown, *Journalism in Our Time* [March 21, 1933], p. 16).
121. Provincial newspapers in England insisted on government control of the telegraph and of radio whereas in the United States insistence on freedom of speech supported private control of broadcasting.
122. *Propaganda by Short Wave*, ed. H. L. Childs and J. B. Whitton (Princeton, N.J., 1943), p. 32.
123. J. K. Winkler, *W. R. Hearst* (New York, 1928), p. 303.
124. Brock, *Meddlers*, p. 116.
125. Silas Bent, *Ballyhoo, the Voice of the Press* (New York, 1927).
126. A. V. Dicey, *Lectures on the Relation between Law and Public Opinion in England during the Nineteenth Century* (London, 1930), p. 302.
127. *The Economic Interpretation of History*, p. 250.
128. Henry James Maine, *Popular Government* (London, 1885), p. 58.
129. *In Defence of the West* (Durham, 1945), p. 118.
130. Cited in G. C. Lewis, *An Essay on the Influence of Authority in Matters of Opinion* (London, 1849), p. 164.
131. Ibid., p. 162.
132. Edmund Blunden, *Keats's Publisher: A Memoir of John Taylor (1781–1864)* (London, 1936), p. 136.
133. Mary Paley Marshall, *What I Remember* (Cambridge, 1947), p. 22.
134. Sir Henry Sumner Maine, *Ancient Law* (London, 1906), p. 320.
135. T. V. Smith, *The Democratic Tradition in America* (New York, 1941), p. 60.
136. Ibid., p. 64.
137. Ibid., p. 63.
138. Walter Bagehot and R. H. Hutton, *Literary Studies* (London, 1879), pp. 374–75.
139. E. J. Urwick, *A Philosophy of Social Progress* (London, 1912), p. 5.
140. Mary Paley Marshall, *What I Remember*, p. 19.
141. Henry Hallam, *Introduction to the Literature of Europe, in the Fifteenth, Sixteenth, and Seventeenth Centuries* (New York, 1887), III, 63–64.
142. Smith, *The Democratic Tradition in America*, p. 63.
143. Gerard de Gre, *Society and Ideology* (New York, 1943), p. 27.

CHAPTER FIVE

Great Britain, the United States, and Canada

Canadians have reason to remember industrial cities in the Midlands for their protests against the imposition of a protective tariff in Canada in 1858 and later dates, following the introduction of free trade in England in the forties. Free trade was accompanied by factory legislation at home and by protective tariffs in the colonies. Thorold Rogers wrote that "a protective tariff is to all intents and purposes an act of war,"[1] and its introduction in Canada undoubtedly appeared to Nottingham and other cities, as an act of war on the part of the colonies against the mother country. The complaints led Canadians such as A. T. Galt to present arguments showing that the protective tariff was not an act of war but was adapted to the demands of a new country and that it was a fiscal device by which improvements in navigation and transportation could be financed, and the cost of moving industrial goods from Great Britain to new markets, and raw materials to Great Britain, could be lowered. British investors were thus insured of a return on capital loans. According to Galt, "The fiscal policy of Canada has invariably been governed by considerations of the amount of revenue required." Moreover, he insisted, "Self government would be utterly annihilated if the views of the imperial government were to be preferred to those of the people of Canada." But the arguments probably made little impression on England. Robert Lowe is stated to have said to Lord Dufferin following his appointment as Governor-General of Canada in 1872, "Now you ought to make it your business to get rid of the Dominion."[2]

I

Throughout the history of Canada, the St. Lawrence River has served as an outlet from the heart of the continent for staple products and as an entrance for manufactured products from Europe. Consistently, political and economic considerations have directed its improvement by the construction of canals and the building of railways. The constitution of Canada, as it appears on the statute book of the British Parliament, has been designed to secure capital for the improvement of navigation and transportation. Railways have been extended from the St. Lawrence to the Atlantic and to the Pacific, and canals have been deepened as a means of increasing the commercial importance of the river. Reliance on the tariff in the Galt tradition has become a crude instrument in the use of which there has been some waste, particularly in duplication of railways, and constant friction over the adjustment of the burden, evident in controversies about freight rates and subsidies to provinces.

To an important extent the emphasis has been on the development of an east-west system with particular reference to exports of wheat and other agricultural products to Great Britain and Europe. However, since the turn of the century, the United States has had an increasing influence on this structure. The construction of the Panama Canal, through the energetic efforts of Theodore Roosevelt, has been followed by the development of Vancouver as a port competitive with Montreal and by a weakening of the importance of the St. Lawrence.[3] The exhaustion of important industrial raw materials in the United States has been followed by the growth of the mineral industry and of the pulp and paper industry in Canada. The Precambrian Shield, which has been a handicap to a system built up in relation to Europe, has become a great advantage as a centre for the development of hydroelectric power and for the growth of a pulp and paper and of a mineral industry in relation to the United States. American imperialism has replaced and exploited British imperialism. It has been accompanied by a complexity of tariffs and exchange controls and a restriction of markets, with the result that Canada has been compelled to concentrate on exports with the most favourable outlets. Newsprint production in Canada is encouraged, with the result that advertising and in turn industry are stimulated in the United States, and it becomes more difficult for Canada to compete in industries other than those in which she has a distinct advantage. Increased supplies of newsprint accentuate an emphasis on sensational news. As it has been succinctly put, world peace would be bad for the pulp and paper industry.

II

The dangers to Canada have been increased by the disturbances to the Canadian constitutional structure which have followed the rise of new industries developed in special relation to the American market, and to imperial markets notably for the products of American branch plants. The difficulties have been evident in the central provinces, Ontario and Quebec, and in provinces which continue to be largely concerned with the British market. A division has emerged between the attitude of provinces which have been particularly fortunate in the possession of natural resources in which the American market is interested and that of provinces more largely dependent on European markets. This division has been capitalized on by the politicians of the respective provinces and by those of the federal government. American branch factories, exploiting nationalism and imperialism in Canada, were in part responsible for agitation in regions exploited by the central area and for regional controversies.

The strains imposed on a constitution specially designed for an economy built up in relation to Great Britain and Europe have been evident in the emergence of regionalism, particularly in western Canada where natural resources were returned to the provinces in 1931, and in regional parties such as Social Credit in Alberta and the C.C.F. in Saskatchewan. In regions bearing the burden of heavy fixed charges and dependent on staples which fluctuate widely in yield and price, political activity became more intense. Relief was obtained by political pressure. A less kindly critic might say that currents of hot air flowed upwards from regions with sharp fluctuations in income. Regional parties have gained from the prestige which attaches to new developments. They have arisen in part to meet the demands of regional advertising, which in turn accentuates regionalism. They have also enjoyed the prestige which attaches to ideas imported from Great Britain, notably in the case of Social Credit and of socialism. The achievement of Canadian autonomy has, then, been accompanied by outbursts of regional activity. Small groups have emerged to combine, disband, and re-combine in relation to protests against the central provinces, notably in the matter of railway rates. Large parties have found it extremely difficult to maintain an effective footing and have tended to break up into provincial parties or into small back-scratching, log-rolling groups within the party.

Provincial regional parties have been in part also a reflection of the influence of new techniques in communication. The radio station, the loud speaker, and the phonograph record enormously increase the power of the regional politician. The radio, for instance, proved a great advantage to skillful preachers in the political field in both Alberta and Saskatchewan. In Alberta, with its vast potential resources, the late William Aberhart, during the period of severe depression and drought, built up a large audience throughout the province using this medium. The influence of Social Credit in Saskatchewan is said to have varied directly with distance from Alberta, the strength of receiving sets, and the power of broadcasting stations. Its success warrants detailed consideration since it points to the elements responsible for the break-up of large political parties. As a teacher Aberhart had acquired an extensive vocabulary. Graduates from his school were scattered throughout the province and his influence persisted as a factor facilitating effective appeal. His Bible Institute and appeals to the Bible and to religion were used with great effect. Bible texts and hymns and semi-biblical language were designed to attack usury, interest, and debts. The conversations and parables of the founder of Christianity were repeated with great skill, notably in attack on the money changers. Audiences throughout the province were held together by correspondence. Large numbers wrote in and subscribed small amounts. Their names were read over the radio and comments were made on their letters. There were attacks on older types of communication such as the chain newspapers dominated by Eastern control. The Calgary *Albertan* was purchased as a means of carrying these attacks into the newspaper field itself.

In the East, Nova Scotia had regarded Confederation as a device for opening American markets, whereas the St. Lawrence region thought of it as a basis of protection against American goods. The Maritimes felt the full impact of capitalism in the destruction of wooden shipbuilding and in expensive transportation to central Canada. Their iron and steel and coal industries, developed to answer the demand for rails and the needs of industrial expansion in Canada, were among the first to feel the effect of a decline in the rate of that expansion. With strong political traditions, born of a maritime background, it might be expected that the Maritimes would be among the first to voice complaint against injustice. Newfoundland has entered Confederation with a great instrument for political intrigue in the federal system, namely admission without responsible government.[4]

The appearance of a large number of small parties in Canada suggests an obvious incapacity of a party or of two parties to represent effectively the increasing number of diverging interests. Provincial boundaries have become important considerations in determining party growth: to mention Social Credit in Alberta, C.C.F. in Saskatchewan, coalitions in British Columbia and Manitoba, Liberals in the Maritime Provinces, Mr. Frost in Ontario, and Mr. Duplessis in Quebec. The consequent complexity suggests a new type of politics or the disappearance of an old type of politics.

The effects of this complexity have been evident in the federal field. At one time government was said to be determined by the longevity of the Walpole administration. The length of life of one administration became an argument for the greater length of life of another. As evidence of the futility of political discussion in Canada, there were Liberals who deplored the activities of the federal administration in no uncertain terms but always concluded with what was to them an unanswerable argument—"What is the alternative?" In one's weaker moments the answer does appear conclusive, but what a comment on political life, that no one should vote against the administration for fear of worse evils to come! One forgets that it probably matters little how one votes so long as one votes against the government or for the party one expects to see defeated in order to secure a healthy minority. All this is in part a result of the exhaustion which accompanies a long term in office, particularly in a trying period, and in the demands of provincial politics. A distinguished federal civil servant once told me that no administration should be in office more than five years. At the end of that time members have ceased to have new ideas or at least are not expected to have any ideas. The exhaustion becomes evident not only among members of the administration but also in the body politic generally.

A further evidence of political lethargy has appeared in an infinite capacity for self-congratulation. Invariably we remark on the superiority of Canadian institutions, Canadian character, and Canadians generally, over Americans. This, of course, is our common North American heritage but in Canada it appears to lead to little more than a congenital tendency toward long arms with which we can slap our own backs. It is a commonplace, of course, that we are encouraged in this by our polite friends from the United States and Great Britain.

III

Our constitution has proved inadequate in the face of the demands made upon it. The Senate, that unique institution, has lent itself to political manipulation. As a guarantee of maritime rights the Maritime provinces were given a substantial number of senators. They have supported the growth of a strong party organization. Politicians have before them as their reward for activity an appointment to the Senate for life. The active part of a politician's life is guaranteed to the party by postponement of appointment. It may be that the Liberal Party will fear an eventual revival of political life and appoint senators who are younger in age so that in case of a political reverse the Senate will continue to be filled for a reasonable length of time with senators loyal to the cause. A careful medical check could be made of senatorial possibilities; the late W. L. Mackenzie King favoured only a general convention that an appointee must be under seventy and have fought an election.

The relation of the senate to party organization has been inadequately studied. The Senate not only provides a useful anchorage for the Liberal Party in the Maritimes but also a support to party organizations throughout Canada. A federal party organizer can be appointed to the Senate and the cost of secretarial expenses charged to services to the country. The procedure has disadvantages in that once senators are appointed they may lose interest in party work since they cannot easily be dislodged, but another senator can be appointed and may bring in new blood. A senatorship is also a reward for journalists[5] who have been active in the party's interest and who will presumably continue active after their appointment. A senator stands as a guard over the party's interest and is expected to be continually alert to the improvement of the party's position in the region from which he is appointed. The entrenched position of the party in the Senate contributes to inflexible government, makes political instruments less sensitive to economic demands, and possibly contributes to the rise of new provincial parties.

Parties are held together to an important extent by patronage and the judicious (not a pun) use of patronage. For the legal profession there remains control over appointment to the bench. Ample salaries, security, retiring arrangements, and prestige tend to make the judiciary a preferred alternative to the Senate. The legal profession and to some extent the medical and other professions are handicapped by professional ethics which prohibit advertising, and the political field is admirably designed to offset this handicap. Lawyers presumably are expected to be concerned with law and it seems eminently fitting that lawyers should be selected by the party to run as members. The substantial advertising developed during the course of a campaign may be followed by the most coveted of all political positions, that of a defeated candidate. The lawyer will not be forgotten by the party when it becomes necessary for the government to select individuals to handle the enormous amount of its legal business. The position of the legal profession in and out of Parliament provides great opportunities for the distribution of patronage. Lawyers will do well, however, to support the party discreetly and strongly since a fanatical loyalty may weaken their prospects of appointment to the bench.

The lack of industrialism in French Canada has meant an emphasis on the church and the law. "Of all the roman provincials the French have been the ones who inherited most of that organizing capacity of the Romans." "It was the French culture of the English ruling caste that made England's power possible."⁶ British governors took over the French bureaucratic administration after the conquest of New France and installed members of the English aristocracy in the civil service. The struggle for responsible government was essentially a struggle for jobs for the native born, a struggle which still continues in Ottawa in the interest of positions for French Canadians in the civil service. To an important extent the history of Canada has been that of a struggle between French and English, and the struggle over patronage has been particularly intense in the legal profession in Quebec.

The importance of the legal profession to party strength necessitates discussion of the calibre of men attracted to it and of legal education, which is hampered in Canada by the broad division of common law and code law, and more seriously by divisions between the universities and educational institutions controlled by the profession, particularly the bench. It is difficult to build up great law schools such as are to be found in the United States, Great Britain, or even Australia. Great legal philosophers have been conspicuously absent. Appointments to important positions such as the deanship of a law faculty have been determined by political prejudices. Consequently the legal profession has lacked confidence and there has been reluctance to take final measures for abolishing appeals to the Privy Council. A strong supreme court is essential to the effective operation of written constitutions, but this has proved to be difficult to obtain, partly because of the necessity of appealing to the Privy Council and partly also because of the handicaps imposed by the British North America Act on systems of legal education through placing education under the provinces.

As a result of its lack of prestige political parties have been able to exploit the legal profession in a fashion which has been the subject of much discussion in legal literature. Legal patronage has been described as "injurious to the independence of both bench and bar." Members of the Supreme Court have been selected to act on royal commissions on subjects in which the government finds itself in an embarrassing position, such as the Hong Kong investigation, the Halifax investigation, and the Communist trials. This use of members of the Supreme Court has fortunately not always met with success, to cite only Mr. Drew's attacks on the Hong Kong investigation, and the failures to secure conviction in the spy trials. Embarrassment to the rights of Canadian citizens has been obvious. A Canadian citizen whose rights may be imperilled by the report of a royal commission which includes members of the Supreme Court will not feel happy about the prospect of appearing before the Supreme Court in a possible appeal from lower courts. The citizen's rights against police interference have been seriously weakened. The use of the legal profession to whitewash political activities of the government is only possible in a country in which the profession has suffered in prestige. The Supreme Court ought not to be in a position in which the government can use it as a doormat on which to wipe its muddy feet.

The lowering of the prestige of the legal profession has implied a heightening of the prestige of the academic profession, with unhappy results for both. The tradition begins perhaps with the late Prime Minister, W. L. Mackenzie King, who came into his position armed with that great academic weapon, a doctorate from Harvard. It would be tedious to trace the steps by which various parties have enlisted the prestige of the academic profession but we can note that members of it were employed on a large scale during the depression, conspicuously with the appointment of the late Norman Rogers as Minister of Labour, and that the trend reached a great climax in the report of the Sirois Royal Commission and in the great trek of the academic profession to the Ottawa salient during the war. Royal commissions have become a device for exploiting the finality characteristic of academic pronouncements as well as of legal statements. The Sirois Report with its length and the number of its appendices was calculated to bring to a focus all the light and leading of the legal and the academic professions in order to produce the great solution to the Canadian problem, and to guarantee the life of the Liberal administration in Ottawa for an indefinite period. It has been used with devastating effect to divide what are called the have-nots and the haves among the provinces and to strengthen the Liberal Party in English-speaking regions. The use of the class struggle as an instrument of politics has been developed to a high point and we could possibly show the Russians a few details in the higher dialectics. The other parties have been paralysed by a situation in which large numbers of voters support Liberals in the federal government, notably in Ontario and Quebec, and at the same time another party in the provincial government, in order that the dominance of any one group may be checked and that a strong opposition may be maintained against the bureaucracy.

During the war period large numbers of the academic profession joined the civil service. Government became extremely complex and the academic profession thrives on complexity. Complexity was suited to patronage, particularly after the war. We may well be concerned with the change in the attitude toward government in Ottawa, since general appeals are made to it for the solution of every conceivable problem, reflecting a belief that governments are omnipotent. We are again thrown back on the limitations of the legal profession in that legislation itself has been used to an enormous extent to strengthen the position of the party and to extend the one-party system in the federal administration.

IV

Heinrich Bruening, former Chancellor of Germany has described basic changes in government and their causes.

> I think that the greatest hindrance to constructive political action in the last thirty years has been the influence on final decisions of experts, especially of experts obsessed with the belief that their own generation has gained a vantage point unprecedented in history. No quality is more important in a political leader than awareness of the accumulated wisdom and experience handed down not only in written documents but also by word of mouth from generation to generation in practical diplomatic, administrative and legislative work. ... The more we work with mass statistics and large schemes the more we are in danger of neglecting the dignity and value of the human individual and losing sight of life as a whole.[7]

Increasing centralization and control by federal civil servants, which have accompanied political difficulties, explain the violation of British traditions of the civil service by which civil servants make pronouncements which are perhaps taken more seriously than those by members of the cabinet. During the war new civil servants, unaccustomed to these traditions, were apparently encouraged to abandon anonymity and to draw fire away from the government. Such pronouncements have been made in the field of foreign policy and reflect the increasing influence of conventions of the United States particularly as centralization facilitates co-operation or collaboration with that country.

The emergence of the civil service to authoritarian control or, to use the German expression, development of *Gruppenführen* and *Ubergruppenführen* has had an important influence on politics. The press is compelled to change its attitude in the news since the facts of governmental intervention are inconceivably dull. Nor is the dullness alleviated by the unrelieved monotony of photographs. Complexity compels the press to emphasize nonsensical subjects or to retreat to issues of the utmost simplicity. The hypothesis may be suggested that the tendency has also made for mediocrity in political leadership. It would be interesting to learn whether calculated stupidity has become a great political asset, but a careful study of the political leaders of Canadian parties leaves little doubt of the existence of the appearance and of the reality. Perhaps political talent is inadequate to the demands of a large number of parties. In any case it would be difficult to find greater political ineptitude than exists in Canadian parties. I must ask to be excused from giving specific examples. Cabinet making becomes "a thoroughly unpleasant and discreditable business in which merit is disregarded, loyal service is without value, influence is the most important factor, and geography and religion are important secondary considerations." Sir John A. Macdonald regarded the ideal cabinet as one over which he held incriminating documents such as might place each member in the penitentiary. Broderick referred to the "malicious credulity of Canadian party spirit and the extreme lengths to which party warfare is carried at the instigation of a most virulent and unscrupulous press."[8] "Comprehensive representation . . . has deprived and will continue to deprive the Dominion of the possible maximum of efficiency in its growing bodies."[9] The demands of the present century have contributed to the exhaustion of political capacity. During World War II the conscription issue destroyed the Liberal Party in Ontario since Mr. Hepburn, the leader of the provincial Liberals, was compelled to oppose Mr. King in the hope of securing Conservative votes. In Quebec the provincial Liberal Party was destroyed by supporting conscription. Political parties have become bankrupt in regionalism.[10]

Provincial parties, or in the words of Professor C. B. Macpherson quasi-parties, hampered in the federal field, have been compelled to undertake measures in their respective provinces which are unacceptable to the federal government. Disallowance of provincial legislation has been a measure of the political necessity felt by the provinces to intensify friction between themselves and the federal government. The difficulties of the British North America Act have been met over a long period by appeals to the Judicial Committee of the Privy Council. The British North America Act has produced its own group of idolators and much has been done to interpret the views and sayings of the fathers of Confederation in a substantial body of patristic literature. But though interpretations of decisions of the Privy Council have been subjected to intensive study and complaints have been made about their inconsistency, inconsistencies have implied flexibility and have offset the dangers of rigidity characteristic of written constitutions.

V

The change from British imperialism to American imperialism has been accompanied by friction and a vast realignment of the Canadian system. American imperialism lacked the skill and experience of British imperialism and became the occasion for much bitterness. American foreign policy has been based on conditions described by Mahan, who quoted the advice of a member of Congress to a newly elected colleague, "to avoid service on a fancy committee like that of foreign affairs if he wished to retain his hold upon his constituents because they cared nothing about international questions." In the Alaskan boundary dispute Canadians felt that they had been exploited by the United States and Great Britain, with results that were shown in the emphatic rejection of the reciprocity proposals of the United States in 1911. But the tide had turned to the point where even those gestures against the United States operated to the advantage of American capital. Branch plants of American industries were built in Canada in order to take advantage of the Canadian-European system and British imperialism.[11] As part of her east-west programme, Canada had built up a series of imperial preferential arrangements in which Great Britain had felt compelled to acquiesce and which proved enormously advantageous to American branch plants. Paradoxically, the stoutest defenders of the Canadian tariff against the United States were the representatives of American capital investors. Canadian nationalism was systematically encouraged and exploited by American capital. Canada moved from colony to nation to colony.

The impact of American imperialism was eventually felt by Great Britain. It began with the spread of American journalism in the latter part of the nineteenth century, and continued notably in the campaign of R. D. Blumenfeld and Lord Beaverbrook in the *Daily Express* for British imperial preference. The campaign was supported with great vigour by the late Viscount Bennett when he became Prime Minister of Canada, and ended in a compromise, as British resistance was gradually mobilized and stiffened.

Participation of the United States in the First and Second World Wars has greatly increased the power of American imperialism and given it a dominant position in the Western World. The shift of Canadian interest towards the United States and the influence of this on Great Britain were brought out sharply in the work of the Right Honourable Arthur Meighen, then Prime Minister of Canada, in persuading Great Britain to abandon the Anglo-Japanese alliance. Canada has had no alternative but to serve as an instrument of British imperialism and then of American imperialism. With British imperialism, she had the advantage of understanding a foreign policy which was consistent over long periods and of guidance in relation to that policy. As she has come increasingly under the influence of the United States, she has become increasingly autonomous in relation to the British empire. Her recently acquired autonomy, marked conspicuously in the first instances by the signing of the Halibut Treaty, has left her with little time in which to develop a mature foreign policy, with the result that she has necessarily felt the effects of the vacillating and ill-informed policy of the United States.

Autonomy following the Statute of Westminster has been a device by which we can co-operate with the United States as we formerly did with

Great Britain. Indeed the change has been most striking. We complained bitterly of Great Britain in the Minto affair, the Naval Bill, and the like, but no questions are asked as to the implications of joint defence schemes with the United States or as to the truth of rumours that Americans are establishing bases in northern Canada, carrying out naval operations in Canadian waters, arranging for joint establishment of weather stations, and contributing to research from funds allocated to the armed forces of the United States under the direction of joint co-operative organizations.

The ease with which such co-operation is carried out is explained in part by the opposition to socialistic trends in Great Britain. Central and eastern, in contrast with western, Canada have had essentially counter-revolutionary traditions, represented by the United Empire Loyalists and by the church in French Canada, which escaped the influences of the French Revolution. A counter-revolutionary tradition is not sympathetic to socialistic tendencies and is favourable to the emphasis on private enterprise which characterizes the United States. Opposition to socialistic devices has been particularly important because large sectors of Canadian economic life have come under government ownership, notably the Hydro-Electric Power Commission of Ontario and the Canadian National Railways. Indeed the large-scale continental type of business organization in private enterprise reflects the influence of governmental administration, in its emphasis on seniority rules and the general sterility of bureaucratic development. Large administrative bodies are compelled to recognize the importance of morale as essential to efficiency. Mobility within the hierarchy can be achieved only with an enormous outlay of energy devoted to the appraisal of capacity. A large number of private enterprises and organizations assume constant attention to the capacities of individuals and are stoutly opposed to the restriction of choice involved in the expansion of large-scale organization. Their concern with private enterprise is reinforced by the views of American branch plants and facilitates American domination.

The abolition of titles has perhaps reflected American influence. The remarks of E. L. Godkin, a native of the north of Ireland, "the most intellectual among American journalists," have been to the point.

> To a certain class of Canadians, who enjoy more frequent opportunities than the inhabitants of the other great colonies of renewing or fortifying their love of the competition of English social life, and of the marks of success in it, the court, as the fountain of honour, apart from all political significance, is an object of almost fierce interest. In England itself the signs of social distinction are not so much prized. This kind of Canadian is, in fact, apt to be rather more of

an Englishman than the Englishman himself in all these things. He imitates and cultivates English usages with a passion which takes no account of the restrictions of time or place. It is 'the thing' too in Canadian society, as in the American colony in Paris, to be much disgusted by the 'low Americans' who invade the Dominion in summer, and to feel that even the swells of New York and Boston could achieve much improvement in their manners by faithful observance of the doings in the Toronto and Ottawa drawing rooms.[12]

"There is nothing in the universe lower than the colonial snob who apes the English gentleman." "These fellows are the veriest flunkies on earth; they are always spouting loyalty and scrambling for small titles and all the crumbs that fall to them from the tables of the aristocracy" (Goldwin Smith). The weakening of the position of these symbols, unfortunate as their effects may have been, has not been without implications for American influence.

American imperialism has been described as "latent and fundamentally political." It has been made plausible and attractive in part by the insistence that it is not imperialistic. Imperialism which is not imperialistic has been particularly effective in Canada with its difficulty in dealing precisely and directly with foreign problems because of division between French and English.

A commercial society in a newspaper civilization is profoundly influenced by the type of news which makes for wider circulation of newspapers—"For God, for country and for circulation." Advertising, particularly department store advertising, primarily demands circulation. Circulation becomes largely dependent on the instability of news and instability becomes dangerous. Effective journalists are those most sensitive to emotional instability. Lack of continuity in news is the inevitable result of dependence on advertisements for the sale of goods. The influence of advertising in the United States spread to Europe, notably to Germany, before the First World War. Bertrand Russell has said with much truth that "the whole modern technique of government in all its worst aspects is derived from advertising."[13] "The intellectual level of propaganda is that of the lowest common denominator among the public. Appeal to reason and you appeal to about four per cent of the human race."[14] "*You cannot aim too low.* The story you present cannot be too stupid. It is not only impossible to exaggerate—it in itself requires a trained publicist to form any idea of—the idiocy of the public."[15] The radio has tended to dominate the news presented in the newspaper, selecting spot news and compelling the newspapers to write it up at greater length because of the feeling that people will wish to know more about the items even though they are not news.

American foreign policy has been to a large extent determined by domestic politics. Publishers of newspapers were rewarded in the patronage system with appointments to ambassadorial posts. The secretary of state has generally played an active role in party politics. An attempt under the second Roosevelt to establish a bi-partisan basis for foreign policy has given greater stability, but foreign issues are all too apt to be dominated by the immediate exigencies of party politics. Under these circumstances a consistent foreign policy becomes impossible and military domination of foreign policy inevitable. The limitations of American foreign policy are largely a result of its lack of tradition and continuity and its consequent emphasis on displays of military strength.

Partly because of the instability[16] of its political system, the United States has shown considerable partiality for generals as presidents throughout its history. The sword has been mightier than the pen, to cite only the defeat of Greeley by Grant. Even in the United States there have been complaints of the pervasive influence of the armed forces, but no signs of abatement are in evidence. Conscription implies a strengthening of their influence. George Ticknor, an American writing in the latter part of the nineteenth century, stated: "Nothing tends to make war more savage than this cruel, forced service, which the soldier who survives it yet claims at last as his great glory because he cannot afford to suffer so much and get no honour for it. It is a splendid sort of barbarism that is thus promoted, but it is barbarism after all; for it tends more and more to make the military character predominate over the civil."[17] De Tocqueville described military glory as a scourge more formidable to republics than all other evils combined. An American has described Washington as becoming the centre for those impelled by the power rather than the profit motive. Bureaucracy assumes a hierarchy, and thus the problem of power.

Formerly it required time to influence public opinion in favour of war. We have now reached the position in which opinion is systematically aroused and kept near boiling point. Strong vested interests in disagreement overwhelm concern for agreement. With control by military men and the difficulties of a Constitution which places power in the hands of the public it may become difficult to check the swings of public opinion. American "candour, good temper, immediate and fearless experimentation, sense for fact, etc., is the positive role of their incapacity for discussions and ideas." "Any fact interests them and *no idea* except as it can be shown to be in direct relation to fact" (Lowes Dickinson). The United States has been described by John Gunther as "the greatest, craziest, most dangerous, least stable, most spectacular, least grown up and most powerful and magnificent nation ever known."[18] Her attitude reminds one of the stories of the fanatic fear of mice shown by elephants.

The Department of Economic Affairs of the United Nations in *A Survey of the Economic Situation and Prospects of Europe* described the trade problem in these words: "The European import-surplus problem is essentially the same as the export-surplus problem of the United States, and the alternatives facing the United States are those facing Europe with the signs reversed; sooner or later the United States must either increase its imports or decrease its exports or do both. But the danger exists that if adequate remedial measures are not taken to work out a tenable balance, the economic structure of both Europe and the United States may become so adjusted to the disequilibrium as to create strong pressures tending to perpetuate it." As suggested by the *Economist* there is a prospect of a "United States dollar shortage forever."[19] Nor does Europe gain much comfort from the United States. Professor J. H. Williams, a judicious observer, writes: "Deepseated in the whole process has been the growing predominance of the United States: resting on the cumulative advantages of size and technological progress and expressing itself in the so-much discussed chronic dollar shortage. . . . We must think of the objectives of the Marshall Plan in terms of reshaping the European economy and adjusting it to its changed world position, and of making the necessary adjustments in our own. We must also regard it as the beginning rather than the end of the adjustment process."[20]

The tariff is an important instrument in American imperialism described in the words of Mr. Dooley as taking up the white man's burden and handing it to the black man. "The mind that thinks in terms of the protectionist symbol is equally at home in the imperialistic symbol."[21] It is as much a contradiction in terms "to speak of protective tariffs as instruments of free enterprise as to speak of militarism or imperialism as instruments of free enterprise."[22] Trade barriers and monopolies become deadly enemies of free enterprise capitalism.[23] Reductions in the American tariff which might widen an outlet for European goods and alleviate the problem have been proposed on a limited scale, but discussion of the tariff in general will not be raised even to the high level of the argument advanced by Mr. Dooley. "The tariff! What difference does it make? Th' foreigner pays th' tax anyhow. He does," said Mr. Dooley, "if he ain't turned back at Castle Garden." There is little prospect of discussion of the tariff in the United States and Canada since European countries cannot expect to have much influence on this subject and, again in the words of Mr. Dooley, "Them that the tariff looks after will look after the tariff."

European countries feel more directly exposed to American influence and to the threat that "the cumulative advantage of size and technological progress" of the United States may enforce uniformity and standardization with disastrous implications for the artistic culture of Europe and for Western civilization. The effects have been evident in the emergence of developments which reflect a profound determination to maintain the supremacy of European culture against the threats of Americanization and communism. Civilization can hardly survive a dumbbell arrangement with its energies drawn to two centres of power, nor an arrangement dominated by one or other power group. Yet it is exceedingly difficult for an Anglo-Saxon trained in a common law tradition to understand the point of view of a European trained in the Roman law tradition.

Canadians can scarcely understand the attitude of hostility of Europeans towards Americans because of the overwhelming influence upon them of American propaganda.[24] Americans are the best propagandists because they are the best advertisers.[25] Whatever hope of continued autonomy Canada may have in the future must depend on her success in withstanding American influence and in assisting the development of a third bloc[26] designed to withstand the pressure of the United States and Russia. But there is little evidence that she is capable of these herculean efforts and much that she will continue to be regarded as an instrument of the United States. The tariff has long since been forgotten in Canada. We too have our mild imperialist ventures, as shown in our acquisition of Newfoundland. "War is self-defence against reform."[27] Neither a nation, nor a commonwealth, nor a civilization can endure in which one half in slavery believes itself free because of a statement in the Bill of Rights,[28] and attempts to enslave the other half which is free. Freedom of the press under the Bill of Rights accentuated the printed tradition, destroyed freedom of speech, and broke the relations with the oral tradition of Europe.

We may dislike American influence, we may develop a Canadian underground movement, but we are compelled to yield to American policy. We may say that democracy has become something which Americans wish to impose upon us because they say that they have it in the United States; we may dislike the assumption of Americans that they have found the one and only way of life—but they have American dollars. It may seem preposterous that North America should attempt to dictate to the cultural centres of Europe, France, Italy, Germany, and Great Britain how they should vote and what education means—but it has American dollars. Yet loans or even gifts are not a basis for friendship. The results are expressed in the remark: "I cannot understand why he is so bitterly opposed to me. I have never done anything for him." Even in the United States a slight appreciation of the definition of gratitude, as a keen sense of favours to come, exists.

In our time we have seen the over-running of Czechoslovakia by Germany with the concurrence of the Allies and on a larger scale the overrunning of Europe in spite of their opposition. But culture and language have proved more powerful than force. In the Anglo-Saxon world we have a new mobilization of force in the United States, with new perils, and all the resources of culture and language of the English-speaking peoples, including those of the United States, will be necessary to resist it. In the crudest terms, military strategy dominated by public opinion would be disastrous.

The future of the West depends on the cultural tenacity of Europe and the extent to which it will refuse to accept dictation from a foreign policy developed in relation to the demands of individuals in North America concerned with re-election. American foreign policy has been a disgraceful illustration of the irresponsibility of a powerful nation which promises little for the future stability of the Western World. In the words of Professor Robert Peers, Canada must call in the Old World to redress the balance of the New, and hope that Great Britain will escape American imperialism as successfully as she herself has escaped British imperialism.

Notes

This chapter is a revision of the twenty-first Cust Foundation Lecture delivered at the University of Nottingham on May 21, 1948.

1. J. E. Thorold Rogers, *The Economic Interpretation of History* (New York, 1888), p. 339.

2. Herbert Paul, *The Life of Froude* (London, 1906), p. 253. "The Canadians, or rather the Maritime Provinces, seem likely to give some trouble, and the British Government may perhaps have an illustration of the difficulties and dangers incident to the retention in diplomatic dependence of communities which are otherwise independent, and which, naturally enough, look to no interest but their own" (Goldwin Smith to Gladstone, May 14, 1871, *A Selection from Goldwin Smith's Correspondence* [Toronto, n.d.], p. 39). In the United States Sumner intended to press "every possible American claim against England, with a view of compelling the cession of Canada to the United States" (*The Education of Henry Adams* [New York, 1931], p. 275). Motley as American ambassador in London in 1870 opposed construction of the Canadian Pacific Railway (Allan Nevins, *Hamilton Fish: The Inner History of the Grant Administration* [New York, 1936], p. 421). "Our relations with England are of far greater importance to us than those with Germany—there being more points at issue, more chances of friction and greater difficulty in almost every question that arises on account of the irresponsibility and exacting temper of Canadian politicians" (Whitelaw Reid to President Roosevelt, June 19, 1906; cited by Royal Cortissoz, *The Life of Whitelaw Reid* [London, 1921], II, 331).

3. See *Canada in Peace and War*, ed. Chester Martin (Toronto, 1941), pp. 58–85.

4. See appendix at the end of this chapter.

5. Mackenzie King prided himself on journalistic appointments to the Senate. Arthur Ford, *As the World Wags On* (Toronto, 1950), 175–76.

6. Wyndham Lewis, *The Art of Being Ruled* (New York, 1926), p. 371.

7. *The Works of the Mind*, ed. R. B. Heywood (Chicago, 1947), pp. 116–17.

8. Hon. G. C. Broderick, *Memoirs and Impressions, 1831–1900* (London, 1900), p. 287.

9. Paul Bilkey, *Persons, Papers and Things* (Toronto, 1940), p. 100.

10. Sir Wilfrid Laurier argued that the caliber of members of the House had declined as business attracted men from politics and law but that the Maritimes continued to send able individuals because of the small character of business there (Arthur Ford, *As the World Wags On*, p. 126).

11. The older type of craft unionism avoided politics and facilitated an international labour movement, but in the newer types of industrial unionism direct intervention in politics in the United States is paralleled by direct intervention in Canada.

12. E. L. Godkin, *Reflections and Comments* (New York, 1895), p. 270.

13. Cited by Denys Thompson, *Voice of Civilization* (London, 1943), p. 180.

14. Ibid., p. 201.

15. Lewis, *The Art of Being Ruled*, p. 91.

16.

> By the Constitution, the Executive may recommend measures which he may think proper, and he may veto those he thinks improper, and it is supposed he may add to these certain indirect influences to affect the actions of Congress. My political education strongly inclines me against a very free use of any of these means by the Executive to control the legislation of the country. As a rule I think that Congress should originate as well as perfect its measures without external bias. (Lincoln in a speech at Pittsburgh, February 15, 1861, in D. A. S. Alexander, *History and Procedure of the House of Representatives* [Boston, 1916], p. 358)

17. George Ticknor, *Life, Letters and Journals of George Ticknor* (Boston, 1880), II, 475.

18. William James referred to "the exclusive worship of the bitch-goddess success." Lloyd Morris, *Postscript to Yesterday* (New York, 1947), p. 330. The vices were "swindling and adroitness, and the indulgence of swindling and adroitness, and cant, and sympathy with cant—natural fruits of that extraordinary idealization of 'success' in the mere outward sense of 'getting there' and getting there on as big a scale as we can, which characterizes our present generation."

19. See F. A. Knox, "The March of Events," *Canadian Banker*, autumn 1948, for a most useful discussion.

20. "The Task of Economic Recovery," *Foreign Affairs*, July 1948, pp. 14–15.

21. E. M. Winslow, *The Pattern of Imperialism: A Study in the Theories of Power* (New York, 1948), p. 203.

22. Ibid., p. 234.

23. Ibid., p. 237.

24. "Says the New York Times' Hansen Baldwin: 'Canada must arm.' " *Time*, Jan. 3, 1949, section on Canada, page 20—an illustration of the crude effrontery of American imperialism.

25. G. S. Viereck, *Spreading Germs of Hate* (New York, 1930), p. 168.
26. See B. S. Keirstead, "Canada at the Crossroads in Foreign Policy," *International Affairs*, spring 1948, pp. 97–110. The problem has not been simplified by the change in the position of the Canadian exchange rate.
27. Emery Neff, *Carlyle and Mill* (New York, 1930), p. 168.
28. See M. L. Ernst, *The First Freedom* (New York, 1946); also O. W. Riegel, *Mobilizing for Chaos: The Story of the New Propaganda* (New Haven, 1934).

Appendix

The following extract from a document presented to the Nova Scotia Royal Commission on Economic Enquiry in 1934 by the late William Rand illustrates the Canadian problem. Federal systems as sources of invective have been largely neglected.

> We use the word Canada in its original and proper sense, comprising the provinces of Quebec and Ontario. It is a common error to regard these two provinces as Upper Canada, and the Maritimes as Lower Canada. That is incorrect. Ontario was Upper Canada, Quebec, Lower Canada. Nova Scotia was never Canada.
>
> At this point we call attention to the lamentable fact that Canadian history as it is written, has suppressed altogether the ulterior facts and motives of Confederation, and garbled these salient incidents which—had they been omitted altogether—would have exposed the historian to ridicule. The majority of these pseudo-historians have been merely the instruments of propaganda to hold back the ever-rising tide of bitterness and resentment of the Maritimes against Canadian violence. The memories of Canadian assault upon the civil liberties of Nova Scotia in 1867 will not down. If the legalized pillaging of our province has constrained its citizens, in self-defence, to drag from dishonoured graves the acts of their betrayers, let the odium rest with the offenders.
>
> In 1867 Canada was a crown colony. Nova Scotia was a crown colony. Surrounded by the sea, nurtured in the traditions of the sea, Nova Scotia became a great maritime power. Her ships were to be found in every port of the world, and they brought back to our people the necessities of life, industry, and commerce, which has made them the richest per head of population of any British colony on this side of the Atlantic. In the colony of Canada, civic strife and hatreds had been assiduously sown. The people were divided in race, language, tradition, and religion. In their Parliament of the two provinces representation was about equal. Every proposal advanced by one province was looked upon by the other as concealing some sinister motive. Suspicion was the watchword of both. The child took it in at the breast, the adult imbibed it and digested it in his home. Legislation at last became a stalemate. The credit of the colony was

gone, both at home and abroad. They were poverty-stricken. Disorder flamed out and Britain became alarmed lest they became involved over the border.

Down by the sea, three maritime provinces, Nova Scotia, New Brunswick, and Prince Edward Island, had devoted themselves to the industries of a maritime people, and had grown wealthy. Nova Scotia had outstripped them all. There were no divisions. They lived in harmony and good-will. It was upon these provinces, and Nova Scotia in particular, that Canada now cast envious eyes. Their plight was desperate. Far outnumbering us in population, if we could be enticed into a confederacy, their domination would be complete, and our revenues and sources of wealth would be at their command.

It is difficult to conceive of your Commission being uninformed of what followed. It is a history of legislative violence inflicted upon a loyal province, a black and evil page in British administration on this continent, in which Nova Scotia was handed over—not to an imperial power to be ruled, not by statesmen and a wise and just sovereign, but to another crown colony, bankrupt alike in morals as in money, and unable to rule themselves. No safeguards, no limitations, were permitted in the wretched betrayal. Spoliation and legalized pillage might stalk in the open, there was no redress. There is none to-day. The Act is without parallel in British history. The "breeds" of the Yukon, the plainsmen of the West, the bushmen of Ontario, the Habitant of Quebec, the bucket shops of Toronto, and the gilded gamblers of St. James St. may issue their edicts to a British people in an island province of the Atlantic, and proclaim to them that they are forbidden to trade in and out of their own harbours under penalty of a pirate's ransom to satiate the inexorable maw of Canada.

Conceive the maritime affairs of Britain dictated by a junta domiciled in mid-Europe of Siberia, and we have Nova Scotia in the Canadian confederacy. Ministers of Marine and Fisheries from the back bush of Canada. Nova Scotia fishermen living within christening distance of the salt sea spray going hat in hand 1500 miles to the back of the continent to get permission to spread a fishing net at his own door.

Delegates from the three Maritimes were named to meet in London in August, 1866. John A. Macdonald, who with Tupper and Tilley, were the prime movers in the conspiracy against the Maritimes, loitered behind till the delegates became incensed and threatened to return home. Upon that Macdonald wrote them:

> It appears to us to be important that the Bill should not be finally settled until just before the meeting of the British Parliament. The measure must be carried por saltum (a leap or a jump) and no echo of it must reverberate through the British province until it becomes law.
>
> If the delegation had been complete in England, and they had prepared the measure in August last, it would have been impossible to keep its provisions secret until next January. There will be few clauses in the measure that will not offend some interest or individual, and its publication would excite a new and fierce agitation on this side of the Atlantic. Even Canada which has been a unit

on the subject of confederation, would be stirred to its depths, if any material alteration were made. The Act once passed and beyond remedy, the people would soon learn to be reconciled to it.

In a century and a half of British history, no more incriminating document has ever been written. No British colony has ever had its constitution so malignantly trampled under foot, or its civil liberties assailed and submerged. If the history of confederation had been written by Macaulay or by Motley, these characters would have been placed where they belong, in the category of political outlaws and ruffians. A century and a half ago, this crime committed in Britain would have sent the perpetrators to the elms of Tyburn or the block. The British Government is not without its share of the guilt. They have attempted apologetically to wash their hands, while at the same time saying to Canada—"Take ye the Maritimes and crucify them, but I find no fault in them."

When it became known that the British North America Act concocted behind closed doors in Downing Street, had been thrust upon Nova Scotia without her knowledge or consent, the anger of her people flamed out. The British flag was torn down and trampled upon. The papers were issued with black borders. The regimental commanders of the Nova Scotia militia—of whom my father was one—waited and prayed for word to come that revolt was afoot. The bayonets of the Halifax garrison were doubled up to intimidate the people. Under these sweet and benign influences Nova Scotia was ushered into a ramshackle confederacy, a confederacy with the same cohesion and unity as the foxes which Samson bound by the tails, each province snarling and pulling in its own direction. The whole British North America Act—whose first sentence is a lie—is but the bastard offspring of the rape of Nova Scotia and the Maritimes by Canada, aided and abetted by Britain. It is within the bounds of propriety and justice to-day, that Nova Scotia should say to Downing Street: "You got us into this hell, now get us out of it."

Our province is degraded and humiliated by constant tramping of delegations to our taskmasters in Ottawa, begging abjectly for that which is our birthright, the birthright of an island province which God gave us, the right to an open and untrammelled sea, to trade, to barter, where we will and can, to bring back to our shores the necessities of life, industry, and commerce, without molestation from an alien power in the back of the continent.

With our loss of population in Nova Scotia, has gone our representation. We are being reduced to the status of a disfranchised negro state of the Southern Union. There are many in Nova Scotia to-day who suggest that we refuse to send representatives to Ottawa and leave the thieves' kitchen to its own. By so doing we would at least maintain something of our dignity. Taking this county of Kings which has been called the garden of the province, its population to-day is exactly where it stood 62 years ago. Its agriculture is decaying. Here are some of the reasons: Mowers which formerly cost $40 now cost $105. Hay rakes, formerly $25, now $50. Plows $10, now $21. Shares 35¢, now 90¢. Disc harrows, $21, now $52. Waggons $85, now $200, and all farm equipment upon the same scale, while within the home the ordinary necessities and decencies of life are assailed with a relentless brutality by Canadian laws which make them impossible. The young man attempting to buy a farm must advance in equipment as much as the cost of the farm. And our people have to listen to the insulting and blatant insolence of the Canadian Manufacturers' Association, telling us how mass production cheapens the cost of the necessities of life and industry.

At the first meeting of the Canadian Senate after Confederation, a Senator from Ontario rose and said, "Nova Scotia has valuable fisheries; we can sell them to the Americans for the use of their railways to the sea." That has been the keynote of Canadian legislation for 60 years. More recently, the Canadian Senate, incapable of a sense of humour or the fitness of things, proposed to take the islands of the British West Indies into the Canadian confederacy, prompted by altruism, as was stated, to help Britain bear the white man's burden. Imagine these hard-headed business men of the southern islands—most of them Englishmen, or sons of Englishmen—the white man's burden, going hat in hand half way to the north pole with their pounds, shillings and pence, like Israelites bearing tribute to Pharaoh, to help blow holes in the shores of Hudson Bay or build free canals for Ontario.

Of all the arrogant edicts issued by Canada against Nova Scotia citizens, that which forbids them buying or bringing home a used car from any American port, smacks the most of the air of the Boston tea party. It was exactly this type of tyrannical edict which drove the American colonies to rebellion. These cars can be bought at half the price of the Canadian, and brought here in our own bottoms at one-fifth the freight from Canada. These edicts are issued at the command of Canadian manufacturing juntas, who hold the government in the hollow of their hand while the Government of our province without spirit enough to defend the people, submitted in abject silence. It is an act of Canadian blockage of our province, pure and simple. It is to no purpose to narrate the putrid story of confederation further.

The Canadian press is the reflex of the Canadian mind. Here are some of the titles by which we are designated: Nova Scotia's Dream Children, The Poor Relations of the Dominion, Another Ireland at our Door. Crucified upon the cross of Canadian ignorance and greed, they have written above us the triple inscription, and the pitiful Canadian mentality calls it wit. Reduced from the richest per head to the poorest of the provinces of the Dominion, jeered by the beneficiaries of the plundering, whose hands drip with the spoils of the Maritimes, and that is Canadian humour.

There is a limit to a long-suffering people's submission to tyranny. Nova Scotia's dream children. Canadian propaganda tells us that as the blind must not dream of sunrise, Nova Scotia must not dream of liberty. To pluck the wild flowers of hope in our hands to warm our hearts, is not for a conquered province, that is the prerogative of a Canadian plutocracy only, for a horde of gamblers whose pawns and stakes are the sweat and blood of the Maritimes, with letters of marque forged in Canada to prey upon the people who live, and go down to the sea in ships. Where in all history did the domination of a maritime people by an inland horde, breed other than strife?

So is it in Nova Scotia today. So will it be. If the blind, stupid, and insatiable greed of Canada, still attempts to inflict upon our people, laws framed with the intelligence of the moron and the instincts of the criminal, we shall defy them.

Index

Aberhart, William, 108
academic profession, 112
Act of Union (Canada), 61
Adams, John, 23–24
Adams, John Quincy, 25
advertising: effects of on writers, 8–9; magazine, 6–8; newspaper, 80–85; political and economic effects of, 90–91; U.S., in Canada, 11–12
Albertan (Calgary), 108
American Mercury (magazine), 7–8
Angell, Norman, 86, 87
arts, versus science, ix–x

Baruch, Bernard, x
Beaverbrook, Baron (William Maxwell Aitken), 78, 87–88, 115
Belford, Alexander, 4
Bell, Moberly, 86
Bennett, Arnold, 79
Bennett, J. G., Jr., 79
Bennett, Viscount Richard Bedford, 115
Blumenfeld, R. D., 78, 81, 115
Boston Watch and Ward Society, 8
British North America Act, 111, 114, 125
Bruening, Heinrich, 113
Bryan, W. J., 83
Bryce, James, 45
Buchanan, James, 27
Burckhardt, Jacob, 2

bureaucracy, Canadian, 64–67
Burlingame, Roger, 9
Butler, Samuel, 83

Canada, 105–25; bureaucracy in, 64–67; Cold War role of, x–xi; culture in, 1–2, 13–14; French, 110–11, 116; imperialism in, 61–62; literature in, 11–12; political changes in, 108–9; postwar, 61–67; relation to Great Britain of, 115–16; relation to Unites States of, xvi–xvii, 2–3, 38, 61–63, 114–21; trade divisions in, 107; transportation and trade in, 106; U.S. media influence in, 11–13, 63
canon law, 56
censorship, 8
Chamberlain, Neville, 88
Chamberlain, Sam, 83
Churchill, Winston, 34, 64, 87
Clark, Champ, 32
Clarke, James, 4
Cleveland, Grover, 29, 31
Coke, Edward, 46
Cold War, x–xi
comic strips, 82
commercialism, 10–14
Committee on Un-American Activities, 61
common law, 46–55, 68n2; and economy, 92; individual character and, 50; present-oriented nature of, 52; Roman law versus, 47, 67, 69n15; social class and, 49

communications: culture versus, xvi; means versus result of, xv; mechanization of, 10–11; and North American colonization, xii; role of media of, 74
Communism, 61, 64, 67
Constitution, U.S., 23, 38–39; as counter to British form of rule, 56–57; weaknesses of, xviii
Coolidge, Calvin, 33, 37, 59
Cooper, Kent, 88
Copland, Douglas, 2
Copyright Act (1891), 3, 5, 6
Crawford, William, 79
culture: Canadian, 1–2, 13–14; communications versus, xvi; effect of mass society on, 10; European versus U.S., 119–21; value of, xviii, 2
cybernetics, xi

Darrow, Clarence, 52, 57, 58
Davis, Jefferson, 27
Democratic Party, 29–30, 32, 34
Dewey, John, xvii
Doubleday, Page and Company, 9
Douglas, Stephen, 27
Dred Scott case, 57
Dreiser, Theodore, 8
Duff, Mountstuart Grant, 86

Easterbrook, Thomas, ix
economics, between the wars, x
Edward VII (King of England), 86
Eliot, T. S., 63
empires, xii–xiii
executive branch, U.S., 23, 28, 31, 33–39
experts, role of in politics, xix–xx, 113

federalism, 67
Federalists, 23–25
Fillmore, Millard, 27
First Amendment, xvi–xvii
foreign policy: conduct of U.S., 34–35, 37, 61; domestic politics and U.S., 117–18; press and, 85–88; weakness of U.S., xviii–xix
France, newspapers in, 88
freedom of speech, xvi–xvii, 120
freedom of the press, 11, 13, 56, 58, 77, 120
French Canada, 110–11, 116
French-speaking Canadians, 13

Galt, A. T., 105
Garfield, James A., 29
Germany, newspapers in, 88–89

Ghandi, Mohandas, 47
Gibbon, Edward, 93
Gibbs, Philip, 80
Globe (Washington), 25
Godkin, E. L., 116–17
Grant, Ulysses S., 28
Great Britain, 45–67; empire and Roman law, 55–56; and foreign policy, 34; legal practice in, 48–55; and Nova Scotia, 123–27; and U.S. imperialism, 115; U.S. politics versus, 34, 39, 43n79

Halibut Treaty, 115
Hamilton, Alexander, 23–24, 56
Harding, Warren, 32
Hardy, Thomas, 11
Harrison, Benjamin, 29
Harrison, William Henry, 26
Harvey, George, 32–33
Havelock, Eric, xiv
Hayes, Rutherford B., 28–29, 57
Hearst newspapers, 5–6, 11, 31, 77, 78, 82, 89
Herbert, A. P., 78
Holmes, Oliver Wendell, 53, 57
Hoover, Herbert, 33, 60
House of Representatives, U.S., 36
Hume, David, 91
Hutchins Commission on Freedom of the Press, xviii
hydro-electric power, 76–77

Ickes, Harold, 61
ideology, in U.S. politics, 28
Innis, Harold, vii–xx
intelligence, short-range versus long-range, xiv–xv
International Paper Company, 75, 77
"Invasion from Mars" (radio show), 89
Ireland, and Home Rule, 47
isolationism, U.S., 60

Jackson, Andrew, 25
Jefferson, Thomas, 24
Johnson, Andrew, 28
judiciary branch, U.S., 56–57

Kent, Frank, 83
Keynes, John Maynard, 92
King, William Lion Mackenzie, 64–65, 110, 112

Ladies' Home Journal, 6
language: and communication in World War II,

89; journalism and literature influence on, 85
law, changing character of, 92–94
law firms, 51
lawyers: education of, 111; and politics, 53–54, 110–11; role in British legislation of, 47–48; specialization of, 51–52; working context of, 48–55
Leacock, Stephen, 2
Lee, Ivy C., 83
Lee, Robert E., 27
Liberal Party (Canada), 109–10, 112, 114
Lincoln, Abraham, 27
Lindbergh, Charles, 81
literature: Canadian, 2; publishing trade effects on, 3–5, 8–10
Lloyd George, David, 54, 87
Loos, Anita, 65
Lovell, J. W., 4
lynching, 71n36

Macdonald, John A., 124–25
MacLeish, Archibald, 61
Madison, James, 24
magazines, 6–8
Maritime Provinces, 108–10, 123–27
Marshall, Alfred, 92
Martial, 1
Massey Commission, xvii
McCarthy, Joseph, 22
McClure, S. S., 6
McKinley, William J., 30–31
McLuhan, Marshall, ix–x
Meighen, Arthur, 115
Mencken, H. L., 7–8, 81, 82
Merz, Charles, 78
Michelson, Charles, 33
military, U.S.: political role of, xix, 37–38; presidency and, xviii, 34, 39; unemployment and, 38
Monroe, James, 24–25
Moore, George, 8
Munro, George, 3–4
Munro, Norman W., 4
Munsey, Frank, 6, 80, 85

Napoleon, 1
National Intelligencer, 25
nationalism, 91–93
newspapers: advertising's effect on, 80–85; circulation of, 5, 84, 117; comic strips in, 82; Continental, 88–89; cost factors for, 75–78; editorials in, 82–83; evening versus morning, 85; feature material in, 82; foreign correspondence in, 85–86; front page of, 83–84; illustrations in, 75, 81; influences on content of, 78–81; and literature, 1–2; syndicates of, 82; tabloid, 81; technological developments for, 75, 94. *See also* press
New York (state), political role of, 37, 43n72
New Yorker (magazine), 7
Niebuhr, Reinhold, vii
Northcliffe, Viscount Alfred Charles William Harmsworth, 79, 80, 82, 85, 86–87, 89, 100n97
Nova Scotia, 123–27

Oppenheim, Phillips, 2
oral tradition: in law, 49–50, 54, 69n15; weakening of, 120
Otis, James, 46
Overthrow Act (U.S.), 57

Parliament: and common law, 46–47; role of, 34
Perkins, M. E., 9
Pierce, Franklin, 27
Pliny, 93
political economy, 92–93
political leaders, 113
political parties, in Canada, 110–14
Polk, James K., 26
postage, role in publishing of, 3
power, intelligence and, xiv–xv
presidents: generals as, 118; succession of, 23–35. *See also* executive branch, U.S.
press: ambassadorial posts for, 84, 117–18; criticism of, 80; role in Great Britain politics of, 87; role in U.S. politics of, 30–31, 33–34; time and, 94; writing of, 79. *See also* newspapers
propaganda, 12, 88–90, 117
publishing trade, 3–5, 8–12
Pulitzer, Joseph, 78, 81, 85

radio: in Canada, 107–8; versus newspapers, 117; political effects of, 33, 34, 42n55, 83–84; in World War II, 89
Ramson, J. C., ix
Rand, William, 123–27
reciprocity treaty (1911), 31, 32, 59, 75, 115
Reed, T. B., 31
regionalism, in Canada, 107–8
Republican Party, 27–28, 31–32

revolutions, political aftereffects of, 21–22
Rogers, Norman, 112
Rogers, Thorold, 73–74, 90
Roman law: British empire influence of, 55–56; common law versus, 47, 67, 69n15; and economy, 92; theoretical nature of, 52–53
Roosevelt, Franklin D., 33, 57–58, 60, 84, 118
Roosevelt, Theodore, 30–31, 37, 57, 59, 75
Roxby, P. H., 2
Russell, Bertrand, 61, 117
Russia, 61, 64

Salisbury, Lord, 80
Sauer, Christopher, 45
Scopes trial, 8
Scribner's, 9
Scripps Howard newspapers, 77, 82, 85
"Seaside Library," 3–4
Seldes, George, 78, 83
Senate, Canadian, 109–10
Sirois Royal Commission, 112
Smart Set (magazine), 7–8
Smith, Adam, 92
Social Credit, 107, 108
socialism, 116
social sciences: common law versus training in, 54–55; and precision, 91–94
South, ix–x
South African War, 86
Speaker of the House, 31–32
specialization, of lawyers, 51–52
Stamp, Josiah Charles, 73
statistics, 91, 93
Statute of Westminster, 115
Steffens, Lincoln, 6
Stephen, James Fitzjames, 66–67
Stephen, Leslie, 80
Storey, Wilbur F., 78
Supreme Court, Canadian, 67
Supreme Court, U.S., 57

tabloids, 81
Taft, William H., 31, 75
tariffs, 105, 119
Tate, Allan, ix
Taylor, Henry, 48

Taylor, Zachary, 25–26
technology: and newspapers, 75, 94; postwar role of, xi
Ticknor, George, 118
Tilden, Samuel, 57
time, concepts of, 94
trade, 119
transportation: and North American colonization, xii; and trade in Canada, 106
Treaty of Versailles, 89, 92
Truman, Harry, 37
truth, facts versus, 58
Tyler, John, 26

United Empire Loyalists, 21, 116
United States: Communist scare in, 61; culture of, xvi–xvii; federal versus state power in, 56; Great Britain's politics versus, 34, 39, 43n79; imperialism of, xvi, 58–60, 114–21; North versus South in politics of, 23, 24, 27–28, 35–36; presidential history of, 23–35; relation to Canada of, xvi–xvii, 2–3, 62; Roman law in, 56
United States Telegraph, 25

"Values Discussion Group" (University of Toronto), ix
Van Buren, Martin, 25, 36
vice presidency, U.S., 36–37

Wallace, Edgar, 2
Wallas, Graham, 73–74
Washington, George, 22–24
The Way of All Flesh (Butler), 83
Weiner, Norbert, xi
Welles, Orson, 89
Whig Party, 26–27
White, William Allen, 80
Wile, F. W., 86–87
Williams, J. H., 119
Wilson, Woodrow, 32, 59, 75
Wolfe, Humbert, 80
women: as consumers, 6; magazines for, 7–8
written tradition, in law, 50–51, 54, 69n15

Zenger, Peter, 56

About the Author

Harold A. Innis (1894–1952) was a Canadian political economist of international renown. After receiving his doctorate at the University of Chicago, he went on to a distinguished academic career at the University of Toronto where he eventually became dean of the graduate school. From World War II until his death, he focused his attention on the newly emerging field of communication and media studies. *Changing Concepts of Time*, his last book, is the third in a series that established the history of communications, or media history, as an important field for critical scholarly research.